A NATURALIST IN ALASKA

AMERICAN NATURALISTS SERIES

Farida A. Wiley, *General Editor*

JOHN BURROUGHS' AMERICA
ERNEST THOMPSON SETON'S AMERICA
THEODORE ROOSEVELT'S AMERICA
JOHN AND WILLIAM BARTRAM'S AMERICA
A NATURALIST IN ALASKA
ALEXANDER WILSON'S AMERICA

A Naturalist in Alaska

By ADOLPH MURIE

Field Research Biologist, National Park Service

Illustrated by OLAUS J. MURIE

*Photographs by the author
and Charles J. Ott*

THE DEVIN-ADAIR COMPANY · NEW YORK · 1961

Copyright 1961 by The Devin-Adair Company.

Canadian Agent: Thomas Nelson & Sons, Ltd. Toronto.
Library of Congress Catalog card number: 59-13559
Manufactured in U.S.A.

FOREWORD

My BROTHER, Adolph, and I grew up in Minnesota, on the Red River. And it was *literally* the Red River, where we swam and skated, according to season, paddled our canoe, camped and fished. I look back on those days as something precious—a bit of the original prairie was still there, a piece of woods was still what we called The Wilderness.

So it seemed only natural that when the opportunity came we both found ourselves in Alaska, traveling with dog team in winter, exploring the appealing Arctic. Adolph went back to finish college in Minnesota, then to the University of Michigan for his doctorate, where he later worked in the Museum of Zoology. He investigated the moose of Isle Royale and made wildlife studies in the Maya country of northern Guatemala while with the University of Michigan. Then followed many assignments with the Fish and Wildlife Service and the National Park Service. Such work involved studies of coyotes, elk, bears, and other animals in Wyoming, elk in the Olympics, coyotes in Arizona and the Yellowstone. Then back to Alaska, and to his favorite country, Mount McKinley National Park, where he concentrated on the ecology of the wolf, but also devoted much time to the study of the grizzly bear and a host of other species. His *Wolves of Mt. McKinley* has attracted a great deal of attention among biologists and others interested in the out-of-doors, and requests for this bulletin have come from a number of countries. It is now out of print.

I believe many biologists approve of the methods used in carrying on this diverse investigation. It is true basic research. It means living with the animals, trying to think as they do, establishing an intimate relationship with the creatures that reveals their motivation in all they do. Such intimate, on-the-ground contact with animals, for as long as it takes to get the desired information, leads to an understanding of nature which is desperately lacking in this age of human exploitation of the planet.

What is much needed today is more mutual respect among the exponents of science, philosophy, esthetics, and sociology. Although we are beginning to think in terms of human ecology, it is now time that we recognize all elements of the good life and give them the emphasis they deserve.

I feel very strongly on this subject. Our civilization is now going through a severe strain. We are trying to find our way, those of us who are concerned with it. And to do so, it behooves us to get serenity in order to think and get back to fundamentals for a clearer view into the future. I believe such writing as this gives a view of truth combined with avenues of natural beauty, as a help toward a richer life.

OLAUS J. MURIE

Moose, Wyoming

CONTENTS

Photographs, between pages 52 and 53, 84 and 85

PHOTOGRAPHIC SECTION

❁❁❁❁❁❁❁❁❁❁❁❁❁❁❁❁❁❁❁❁❁❁❁❁❁❁❁❁❁❁❁

ACKNOWLEDGMENTS

I WISH TO THANK the following publications for permission to use material that originally appeared in their pages: *Audubon Magazine, Nature Magazine, The Living Wilderness,* and *National Parks Magazine.* I also wish to thank the National Park Service, and particularly its Director, Conrad L. Wirth, for permission to reprint portions of the *Wolves of Mt. McKinley* and for the encouragement I have received in my work.

I join with the publisher in extending grateful thanks to Alexander B. Adams for his careful editing of the text and his arrangement of the material in this book.

Special thanks are also due Farida A. Wiley, general editor of The American Naturalists Series, for reading the text and making many valuable suggestions.

A.M.

A NATURALIST IN ALASKA

1. Wilderness — North

ALASKA has for most of us a magic ring. It is still a frontier, and chiefly a big wilderness. The freshness of primeval tundra landscape is the magic for true Alaskans. It is a land where the individual is not yet swamped by numbers. The ranges and migratory routes of several herds of caribou have not yet been preempted. Their rights may not have been officially recognized but as yet there have been few claim jumpers. The mountain goats, the white mountain sheep, and the moose still wander freely as in primitive days. Here one may even see the wolverine and hear the mournful music of the northern wolf.

Alaska is not quite the same, of course, as it was in 1920 when my brother, Olaus, traveled by ship to St. Michael on a thrilling assignment and a big one. He was to study the habits of the caribou and map their migratory routes and estimate their numbers. A reindeer industry appeared to be flourishing in the Nome region, and E. W. Nelson, Chief of the old U. S. Biological Survey, who had been a naturalist in Alaska in the late seventies, was worried about the caribou. He did not want the inferior domesticated reindeer brought into the caribou ranges, causing the caribou to become inferior by cross-breeding with them. The chief problem in this regard was to learn what parts of the country were used by caribou, so that the expansion of the reindeer industry could be regulated accordingly. My brother had already spent two summers and a winter in the Hudson Bay region, engaged in natural-history studies, and had traveled through the interior of Labrador from south to north by canoe, so he was familiar with the north country and was a self-sufficient traveler. These were the blessed days before the advent of the airplane in the north. In winter he journeyed alone by dog team, in summer he hiked cross-country, packing his dogs and living on blueberries, ptarmigan, and what other meat he could secure for himself and dogs. The Sourdoughs scattered over the country in those days had a familiarity with wildlife and helped him in many ways. His mere arrival at each occupied isolated cabin was an entree, a "making good," for he had reached there by his own efforts, and the hospitality offered was unbounded. He soon had a reputation in Alaska as a wilderness traveler and it was well deserved, for no one was more tireless, both physically and mentally. This kind of travel led to an intimacy with the tundra. During his journeys he wrote his notes at night and when he found time made drawings of the animals and the country. It is with this background that he has, from time to time, made the sketches for this book, in hours filled with his own writing, drawing, painting, and serving as Director of the Wilderness Society.

Olaus had always been more than a big brother to me, so when, in the summer of 1922, before I had finished college, I had the opportunity of becoming his assistant, it was not only the promise of high adventure but also the anticipation of the companionship that thrilled me. After closing a summer project in Mount McKinley National Park we went to Fairbanks, our headquarters. Here we met professors and students of the just-opened University of Alaska and many old-timers, some of whom became lifelong friends. But we chafed to be off on an all-winter dog-team trip we had planned, a reconnaissance covering the country to the north. Cold weather, twenty to thirty-five degrees below zero, arrived in late October and early November, but the snowfall was too light for sled travel over hummocky tundra until quite late that fall.

We had seven dogs — four which were the nucleus of the team my brother had been using, and three more that we secured from Van Bibber, rangy old-time miner, hunter, trapper, and dog man, who kept his dogs on the banks of the Chena River. The tall, rawboned Van Bibber had such a big, deep voice that, when he thundered the names of his dogs — Irish, Dawson, or Hooch — the names were never forgotten; they still ring in my ears. One of the dogs, a huge, yellowish animal, with some strange outside blood in his veins, with mulelike ears, the runt of the litter when born, weighed 140 pounds. This was to be his first full winter of work, for he was just past puppyhood. We wondered if he might not be too clumsy, but we soon learned that his gait was a tireless pace. His later career was unusual. He was used as a hero dog in a Norman Dawn movie because of his appearance; then he showed up posing on a Charlie Chaplin "Gold Rush" float in California. Still later he was used as an exhibit in an eastern chain of department stores, and a caption under a picture of him in a New York newspaper said he weighed 190 pounds, which he probably did. When he joined the movies his name was changed from Irish to Ilak. But I am sure that his happiest days were his hardest, those spent leading one of our two teams into the Brooks Range.

While we waited for sufficient snow we exercised the dogs on daily runs out to the Tanana River. Before ideal conditions arrived for dog mushing, we shipped dogs and sleds in a freight car to Nenana, and in the dark the next morning we were on the trail with two sleds and seven large dogs, en route for Minto, the first roadhouse. In a few days we reached Tanana, where one of the last Army posts in Alaska was being abandoned; then down the Yukon River to Kokrines, into the hills to a reindeer herd, back to Tanana, and across country to Alatna on the upper Koyukuk River where we arrived in time to take part in Christmas and New Year festivities with the Eskimo and Indians — ballroom and square dancing until two in the morning. The trader, Sam Dubin, had forgotten his false teeth in one of the mail relief cabins and we brought them with us. Johny Tobuk, the Eskimo square-dance caller, hung them on the Christmas tree.

En route we had picked up more dogs until at one time we had thirteen. They were the best dogs we could get, recommended by good dog men. Two of them were quarter-breed wolves. It was an all-star cast, and each was a temperamental star, ready to fight anyone who might look his way, and every dog tried his best to join every fight. Only sharp blows on the nose would stop these daily battles, and we became adept referees, jumping as automatically into a fight as any of the dogs.

From the Alatna River we went across country toward the Kobuk River, but dense willows and alders over the swampy country had slowed travel so much that we had to turn back. Then about 150 miles of trail breaking to the head of the Alatna and Kutuk rivers to secure the first specimens of mountain sheep in the Brooks Range. For shelter we carried a seven-by-nine silk tent and a stove. Temperatures for January and February averaged about thirty-seven below zero and reached sixty-eight below. Yet we managed easily and routinely — our equipment was simple but all that we needed. Then to Wiseman, made famous by Bob Marshall's excellent book, *Arctic Village*. From here it was cross-country to the Chandalar River,

Beaver, Fort Yukon, Circle, and back to Fairbanks, where we arrived on April 26 on the last of the winter snow and found the ground already bare.

Since this wonderful all-winter trip we have both spent much time in Alaska. Olaus mushed dogs one spring from Nenana to the fabulous Hooper Bay bird-nesting grounds, where snowy owls, loons, cranes, geese, ducks, shore birds, jaegers, and many other species gathered to bring forth their young. One spring it was to the Alaska Peninsula to study this treeless region, its bird and mammal life, and particularly to collect some brown-bear specimens. He made two summer expeditions by boat the length of the Aleutian Islands to study the colonies of nesting sea birds and to observe the status of the sea otter that had made a comeback in the area after coming close to extermination there at one time. More recently he headed a summer expedition into the Brooks Range, part of which has since been set aside as the Arctic National Wildlife Range. After a hiatus, my own Alaska work continued. In 1939 I was assigned to study the ecology of the wolf in Mount McKinley National Park, summer and winter, and I have been returning to this and other parts of Alaska at intervals ever since.

The wildlife information contained in the following chapters was chiefly gathered at Mount McKinley National Park, which was set aside in 1917. The idea for national park status no doubt had its inception much earlier in the minds of some of the early visitors. In 1906 and again in 1907 the late Charles Sheldon, hunter-naturalist, worked his way deep into the wilderness of central Alaska. He sought simplicity, solitude, the feel of weather, and a close acquaintance with animals in the remote mountains of the Alaska Range. The cabin he used as a base he built in a patch of woods near the last timber toward

the head of the Toklat River, which has its source in glaciers
lying along the crest of the range. This cabin is now in ruins,
and the cache when I last saw it was tottering. A porcupine
was using one of the several log kennels he had built for the
sled dogs. The Toklat River had washed away the woods up to
the edge of the ruins, leaving them dangerously exposed. But
the gravel bed of the river in front of the cabin is broad, perhaps a half mile across, so that it may be years before the
stream again moves over to where the cabin stands.

During various trips down the Toklat in the course of my
field work, I usually stopped at the cabin, lingered to examine
the walls, the shelves, the wooden pegs used for nails. I would
stand before the cabin and look across the gravel bars to the
mountains, a scene that Sheldon must often have enjoyed. Although the cabin is deteriorating, and a swing of the river may
destroy it suddenly, I have a feeling it should be left alone. I
think that Sheldon, with his love for wild places, would like to
have his cabin crumble to earth with age.

This wilderness which Sheldon knew so well is now a part of
McKinley Park. Largely through the efforts of Sheldon and
James Wickersham, delegate to Congress, the region was finally
made a sanctuary for wildlife. The park stretches more than
one hundred miles along the Alaska Range and is from twenty
to thirty miles in width. Most of it lies to the north of the crest
of the range and includes the foothills and some of the more
level tundra beyond. McKinley, with its 3,030 square miles, is
our second largest park, but it is only a small fraction of the
vast surrounding wilderness.

Lofty Mount McKinley is so remote and grand it hardly
needs protection. It rises higher from its base than any other
mountain in the world, about 18,000 feet, and is the highest
mountain on the North American continent. Even without

this dominating feature, McKinley Park would be outstanding because of its alpine scenery, its arctic vegetation, and its wild-life. I have walked over the green, flowering slopes in the rain, when the fog hid the landscape beyond a few hundred yards, and felt that the white mountain avens, the purple rhododen-drons, and the delicate white bells of the heather at my feet were alone worthy of our efforts.

How often have people looked longingly to that northern corner of our continent, with thoughts of Arctic expeditions, glaciers, dog mushing, and far places! In McKinley Park, a choice portion of Alaskan wilds has been made accessible and, so far, mechanical facilities do not obtrude unduly. It is still possible to get away from camps and roads far enough to feel that you are in Alaska.

The animals all belong; they are original Alaskans. Alaska without caribou or ptarmigan would lack much of its charac-ter. Alaska full of transplanted elk and Chinese pheasants would no longer be Alaska.

Olaus and I cherish the hope that the Brooks Range in northern Alaska, the Alaska Range that passes through Mc-Kinley Park, and the coast ranges, and wide sweeps of inter-vening country, may all be kept forever wild. That the rivers may remain rivers. That the tundra, with thousands of ponds of all sizes, may be left sufficiently intact to serve in the future as nesting homes for the cranes, the many species of shore birds, the ducks, geese, and swans. All this, so that man in the future may continue to enjoy wild country.

In the late 1930's there was widespread concern over the increase of wolves in Alaska. Some efforts had been made to control the wolf, especially in areas occupied by domestic reindeer, and there had been apprehension in some quarters concerning the welfare of the big-game herds elsewhere in Alaska. Since McKinley Park lies in the heart of the Alaska wolf range and carries its proportionate quota of the general wolf population, there had been considerable agitation for wolf control in the park.

Fortunately, there is wise provision of long standing in the policy of the National Park Service that no disturbance of the fauna of any national park shall be made until a proper scientific appraisal of a situation has been made. Consequently, before anyone could embark on a program of wolf control, a number of questions had to be answered. In 1939 I was given the assignment of determining the relationship between the wolf and the Dall sheep. It was virtually a virgin field for scientific investigation. Scarcely anything was known about the wolf's home life or his relationships to mountain sheep, caribou, moose, and other smaller species.

I arrived in the park on April 14 and three days later was taken twenty-two miles into the park by dog team and left at a cabin on Sanctuary River where I started my field work. The next morning I climbed a mountain and saw a ewe and yearling on a grass slope, the first white sheep I had seen in sixteen years, and a little later, through the field glasses, I "picked up" a beautiful ram resting on a ledge, the graceful curved horns silhouetted against a spring-blue sky. A strong, cold wind was blowing on top, and I slipped on my parka. During the day I classified sixty-six sheep, twenty of which were yearlings. I saw wolf tracks and a wolf dropping containing sheep hair. The long, slow process of gathering data had begun.

To avoid accumulating miscellaneous observations only, I directed my efforts along the most promising lines and kept in mind the main points on which it was desirable to get quantitative data.

First it was necessary to learn what the wolves were eating. Killing the wolf to examine the stomach contents, in this case, was too much like killing the goose that laid the golden eggs. A dropping tells almost as much; so, to learn the food habits, I gathered wolf droppings and, of course, made observations of the wolves in action whenever possible. From the analysis of many droppings, I obtained some notion of the extent of the feeding on mountain sheep. The next point to determine was to what extent the sheep eaten represented carrion. It could all, of course, have been carrion. Some of it certainly was. But if I were to learn that sheep are commonly run down and killed by wolves, then it would be necessary to learn what kind of sheep are killed. Were they the ailing, the aged, and the young, or were all classes being taken indiscriminately? These were points difficult to determine. A thorough search of the sheep hills yielded skulls of 829 sheep. These skulls showed in what types of animals the mortality in the population lay and brought up fundamental problems concerning the natural role of the predator.

Another obvious line of attack was to classify the white-sheep population to determine the size of the lamb crop and the survival of yearlings in order to learn the losses during the first year — a critical period. I made classifications in most parts of the range at every opportunity and gathered all other available information on the wolves and sheep. The food habits and range of the sheep were of special interest, for adequate range is often an important factor in predatory problems, and I sought historical data on the wolf and sheep populations, for such data frequently are enlightening.

Involved in this same problem were other animals which required study to determine what part they played in regard to the sheep and to the wolf. The caribou, a basic source of food for the wolf in interior Alaska, had an important bearing on the wolf-sheep relationships. There were such questions as the extent to which the caribou serve as a buffer species for the mountain sheep. The caribou, I studied in much the same manner as the sheep.

It is a tradition that the golden eagle, which shares the high country with the sheep, is one of its principal enemies. Perhaps the golden eagle was levying too heavy tribute on the sheep — or perhaps it was a valuable citizen. Therefore, the food habits and actions of this bird required attention. Another animal on the agenda was the grizzly bear, known to be fond of caribou and sheep meat. The wolf has been accused of being destructive to foxes, so it seemed important to make what observations I could on this point. Furthermore, since foxes are a potential enemy of lambs, their food habits required study. The moose was present in some numbers, and since it is generally considered a source of food for wolves, I gave some attention to this species also.

I had hoped to make observations of the coyote in the sheep hills, but it was too scarce for study. Coyotes are more plentiful in areas adjacent to the park in the rabbit country, a fact which may in itself be significant. Other animals, such as the porcupine, marmot, ground squirrel, snowshoe hare, mouse, and ptarmigan, were considered in their relationships to the larger forms.

Such, in brief, was the scope of the study.

In 1939 the field work, begun in April, continued to the end of October. In 1940 I returned to the park in April and for the next fifteen months remained in the field. The field observations thus were made over a period including most of three spring and summer seasons, two autumns, and one winter. The work necessitated a large amount of hiking and climbing. In 1939 I walked approximately 1,700 miles. The following two summers the work still required much climbing but fewer long hikes. During the winter I traveled on skis, carrying a bedroll and food in most cases and using relief cabins for shelter.

2. The Lynx and the Pendulum

IT WAS YEARS before I saw my first lynx, even though I had spent considerable time in lynx country. It is a cyclic species, and I had witnessed a prolonged scarcity phase in McKinley Park, in which I had not noted even a track. Nevertheless, even when scarce or absent, the lynx was part of the McKinley wilderness, for its scarcity was natural. Scarcity and rarity have their compensations. As the years passed by, the romance of the lynx I had read about in Ernest Thompson Seton's books became ever greater. At long last, after the prolonged scarcity, the lynx returned with an upswing of rabbits in the fifties and stalked with royal dignity through his rabbit domain like a wealthy feudal baron — which he was, for a day. Like civilizations, he has his rise and fall. Tomorrow he would be starving or bruising himself capturing a Dall sheep. Such is fate for the lynx tribe.

In general the lynx expresses little fear of humans. A few jumps into the brush and he stops, or when discovered he may stand watching, seemingly unperturbed. On Thorofare Pass one evening a lynx made half a dozen jumps which took him out of view and into a dry, rocky stream bed leading up a steep slope. When I arrived at the spot where he had jumped away, I looked up the slope for him, failed to see him, and wondered where he could have gone, for there was no cover nearby. Then, from among the rocks where he had stood watching me, twenty paces off, he bounded away. He had blended with the boulders in the evening shade of the bank, so close to me that I had overlooked him. As he loped up the rocky slope he was soon hard to follow with the eye, for his grayness was like the shadowy background. Perhaps part of his tameness is due to his reliance on escaping discovery.

On another occasion I surprised a lynx on a treeless open slope where the vegetation was only a few inches tall. His behavior was different, for he galloped down the slope in long leaps and did not stop until he had reached a narrow fringe of willow brush along a stream a quarter mile or more away. His fright was perhaps due to being startled, and his unusually long run to the lack of cover. Under the circumstances his behavior seemed logical.

Charles Ott, who has had much experience observing and photographing lynx in the park, has been able to follow them for hours in winter without unduly disturbing them. Occasionally a lynx has been approached for pictures as close as necessary. There has been, so far as I know, no record of attack on humans. Behavior among animals is always variable and uncertain, but the lynx, to sum up impressions of him, behaves as though unafraid, indifferent, self-contained, and fully concerned with his private affairs.

The lynx has a lean body, long, thick legs, and broad feet. The hindquarters are so well developed and high that a mystified observer described him as an animal that appeared to be walking downhill when walking on the level. The eyes are startlingly big and yellow. The face has a squarish contour, and the throat ruff is prominent. The ears are tipped with a tuft of long, glossy-black hairs. The stubby tail is about four inches long and tipped with black. The pelage is rather long, grayish in winter, more tawny in summer. There are a few blackish markings on the face and other faint mottlings over the body. A fully adult lynx in good flesh usually weighs from twenty to twenty-five pounds, occasionally a few pounds more, and kittens in their first winter have been recorded as weighing ten to twelve pounds. The height of an adult is apparently in the neighborhood of twenty-four inches. Compared to the bobcat, the lynx has longer legs, larger feet, and longer ear tufts; it lacks the bobcat's more definite black markings over the body; and its tail is black-tipped above and below, while the stubby tail of the bobcat is black only on the upper side of the tip. The hind foot of the lynx measures eight to ten inches, and that of the bobcat usually less than seven inches. The bobcat apparently averages a little heavier.

The range of the lynx lies largely in Alaska and Canada, reaching southward in a narrow strip to Oregon and Colorado in the mountains and into the northern states from Minnesota eastward. For the most part, its range lies north of that of the bobcat. Three subspecies have been recognized, at least one of which is said to be of doubtful status.

When nature decided, so to speak, that the lynx should forevermore, or at least for a long time, be chiefly dependent on the snowshoe hare for his welfare, she seems to have initiated a similar development for survival in both species. To simplify the unknown complications, we might say that the snowshoe rabbit had to develop snowshoes to escape the lynx, and the lynx had to develop the same type of footgear to catch the rabbit. The rabbit, already having strong hind legs, developed a snowshoe type of foot where it would do the most good — on the hind legs. The lynx developed the snowshoe effect on all feet; the forefeet, being a little larger to start with, remained larger. Consequently, relationships remained nip and tuck between them, and both species have prospered so far as their mutual relationships are concerned.

The lynx's track is slightly broader than it is long. In walking or trotting, the imprints consist of four toe pads and a central pad. When galloping or taking a long step, another pad on the front foot, which ordinarily is above the surface, may show. Likewise the heel of the hind foot may show at times, especially in loping or jumping. In wet snow the heel leaves a narrow, well-defined imprint.

On the morning of June 12, 1955, I stopped at a woods near Savage River to inspect some lynx tracks. It had been alternately snowing and melting the past two or three days, and during the night new tracking snow had fallen. While inspecting the tracks and getting ready to follow them, I saw a lynx some thirty or forty yards back in the woods. It had stopped to look at me and now walked in an arc to pass me. After neatly hopping a ditch it broke into a light, springy trot and disappeared in the willow brush that grew in the rather open spruce woods.

While watching the lynx, I had been conscious of an unfamiliar sound up in the woods in the direction the lynx had taken. Perhaps an odd call by a camp robber! Its quality reminded me of the organlike call of the varied thrush. After the lynx disappeared, I listened more attentively to the sound and concluded that it was the cry of some animal in distress. Moving up through the trees, I located the cry under a fallen spruce. I was just in time to make out the form of a lynx under the windfall picking up a small kitten and starting north with it. My immediate conclusion was that the lynx, her young discovered, was moving them. But then I

learned that the rest of the kittens had already been trans-
ported, and now she had gone off with the last one, which
perhaps had been crying because it was lonesome.

I suspect that the move was made because the windfall had
been a poor shelter during the recent wet and snowy weather.
There were dozens of windfalls nearby in the woods that
would have given better protection, and so would many of
the standing spruces that branched densely close to the
ground. The bed most used was inadequately protected from
above by the trunk of the windfall, where it was about sixteen inches in diameter, and by a few limbs. Moss that had
originally been scraped together to form the bed was soaked.
The bed had been much used, for considerable lynx hair
clung to the debris. Later it appeared that twigs had been
placed on the moss, perhaps in an effort to create a drier
bed. I have observed a husky dog rearrange its stick bed, removing the heavier branches to make it more comfortable,
an action somewhat similar to what the lynx had done. But
the bed with the loose twig covering had not been used
much. Another bed nearer the base of the windfall had just
been vacated. It was here that the kitten had been picked up
by the parent. The drawback to this bed, aside from a
meager roof, was a heavy limb lying across the middle of it
which did not leave enough room for comfort.

The kittens were moved about two hundred fifty yards to
a large, bushy spruce windfall that formed a dense canopy
for nearly the entire length of the tree. The branch screen-
ing on the sides was so heavy that the mother lying with the
kits apparently felt secure, for she was not perturbed when
approached to within about twenty feet. She lay on her side,
watching sleepily, frequently letting her eyes close. The
young, which appeared to be ten or twelve inches long, wad-

dled clumsily around and over her body, on unsteady legs. Their eyes were open. They were perhaps not much over two weeks old — probably born the latter part of May. They made a few soft mewing sounds. There were at least three young at this time, but others could have been hidden, for eleven days later I counted six.

I visited the den daily for brief periods the next four or five days. The lynx that I assumed to be the mother was usually with the kittens and continued to be unperturbed. On June 23 she was absent in the morning and I counted six kittens. They were near the tip of the windfall when I first saw them, but then they moved toward the middle where they were better screened.

On the morning of June 24 I approached a group of trees about thirty yards from the lynx home. One of them I could climb without being seen, and from it I had a good view of the windfall. I arranged my packboard for a seat on some limbs about twenty-five feet from the ground and draped a blanket over a limb in front of me. I had not long to wait for activity. At eight-thirty an adult came forth from the far end of the windfall. Sitting on her haunches, she looked briefly in my direction. After finishing a wide yawn, she walked slowly and deliberately off through the woods. The little ones cried a few times. In the direction the lynx had taken, a red squirrel scolded.

Quiet prevailed for an hour and a half. Then a rabbit skittered across an open space as though escaping, and I watched for a pursuer. A few moments later a lynx emerged — not chasing the frightened rabbit but walking with stately, regal steps, looking neither to right nor left but straight ahead, seemingly concentrating on its own lofty thoughts and oblivious of its surroundings. It was a beautiful animal, and I

thought it heavier and much more tawny than the lynx that
had left the den earlier. I wondered if it was the mother's
mate, but this I could not determine — certainly it behaved
in the manner of some royal male. But this independent
creature was not oblivious to its surroundings. When it was
directly opposite me in line of march to the den, the short
thick tail stood straight up. The black tip was conspicuous.
Then *snap* went the tail to one side, and *snap* to the other
side, and thus at deliberate and regular intervals the tail
was snapped from side to side. There was no other indication of emotion. Possibly the lynx had caught my scent, or
was aware that it was passing by my position. After going behind a clump of spruces it came forth near the den. The tail
had stopped jerking. My camera clicked. The lynx halted.
The tail started jerking again and the lynx looked toward
me — a brief look. Then it marched on with the same deliberate step until, with jerking tail, it disappeared under the
windfall — tall glossy-black ear tassels and measured step —
deigning to look aside only once. Never was there a more
proud, self-possessed and impressive stage entry.

For two more hours I watched, but I saw no more of the
lynx family. A pair of pigeon hawks were aroused to cry out
a few times, once when a marsh hawk sailed too near. A pair
of white-crowned sparrows lit on the windfall, then one gave
a few sharp warning chirps and they flew away. Two camp
robbers investigated, as though hopeful of finding food. Off
on the mountain were seventy sheep, gathered in a compact
flock, watching the valley below them for danger, for they
had to cross this lowland to reach their summer ranges.
Hudsonian chickadees foraged about the branches near the
place from which I watched. At noon I was frozen out, still

convalescing from flu, and, anyway, one big thrill was enough for the day.

The following morning I watched from the tree from seven forty-five until eleven-thirty. It was another cloudy, cold day. The mother was stretched out at one of the entrances leading under the windfall, where I could see her quite well. She licked her feet and various parts of her body. Two or three kittens at a time were seen moving about her and they, too, were groomed whenever they came near. The kittens made a few mewing sounds, and twice the mother made a low yowling sound when it seemed that the kittens were bothering her.

The next day, June 26, at nine A.M., I caught a glimpse of the mother slipping away from the tip end of the windfall. Announced by a red squirrel, she returned in the same manner. In the afternoon she again lay partially exposed, stretched on her side, once rolling over on her back, and occupied herself with much grooming. In grooming, she would at times lick a paw and pass it over her face. The young lolled around, much of the time sleeping close to her. She looked my way a few times when she heard the camera shutter, but in a moment she would drop her head to the ground to rest again.

June 27 was an eventful but rather unfortunate day. I was at the lookout tree at seven in the morning. Before I had finished arranging myself in the tree, and before I had draped a blanket over the limb in front of me to serve as a partial blind, the gray lynx came out from the windfall and sat on her haunches facing my way. I cautiously snapped a picture. She apparently heard the shutter, for she looked up. Suddenly she recoiled, drawing back under the branches of

the dead tree behind her, as though she had seen an appari-
tion. She seemed shocked with dread and terror, and in a
few moments she slunk away. Perhaps it was the strangeness
of my unusual position in the tree that scared the lynx. After she left, although I did not see her, I was sure she was
moving nearby, for to one side behind me I heard a squirrel
scold. She was no doubt investigating my presence. An hour
later, from behind and to one side of me, I heard a soft murmuring sound, "m-m-m-m," almost moaning and rather high-
pitched. It was similar to the sound a mother fox uses to call
her pups forth from a burrow. A few moments later the lynx,
still calling, trotted quickly to the den in a crouching position, as though she had made a definite decision.

She came forth from the den at once, calling "m-m-m" as
she slunk away. Some of the kittens tumbled after her and
moved out of view as she continued to call softly. But in a
few minutes the young tumbled back to the den again. She
returned two or three times, calling for them to follow, but
they were loath to leave their home. She returned and all
was quiet for half an hour, and I thought the family was re-
settled. Then, to my surprise, an adult returned to the den
with two or three young. This adult seemed different from
the one I had last seen enter the den. I had the impression
that it was more tan colored, like the one I had seen returning to the den a few days earlier. There seemed to be a very
tan adult and a gray one, but I could not be sure, for I
never saw both at once.

The lynx could leave the tip of the windall without being
seen by me, so I was not sure what might have occurred that
I had missed. Then an adult again departed, followed by
two or three kittens. A young one left behind cried loudly.

Soon I heard young lynxes crying up in the woods. The gray adult returned, trotting rapidly, and she was very intent. She came forth with a kitten that traveled poorly. Perhaps it was the runt. She picked it up in her jaws, dropped it for a better hold, and carried it with difficulty across the open space and out of sight.

The crying up in the woods continued, and I left my tree and walked up in the woods to one side of the place from which the sound was coming. Soon there was crying on both sides of the highway, and I saw a young one waddling back across the road. An adult returned it, literally pushing it across the road with her nose, and then two or three others wobbled across too. They were headed for the original den. I walked back so as not to disturb them and to permit them all to get across the road safely.

I looked around the den for young, but they were all gone, and when I returned to the road I heard no more calling, and the family had disappeared. I never saw the young again. The next two or three days, three of us searched the patch of woods, inspecting dozens of windfalls, but to no avail. The family could have moved out of this woods or could have hidden under one of the many spruce trees or windfalls that we missed. We saw a large tan lynx during our searching.

I have since wondered if two lynxes were at the den, for, as I have pointed out, I had the impression I was seeing a gray and a tawny animal. But the light could have made the difference. However, one visitor at the den stated that he was rather sure that two adults were under the windfall. If there were two adults, was it a pair or two females?

This was the only denning lynx I have encountered. The following summer, on August 14, I found a spruce windfall in the woods in the Igloo Creek area that had served as either a den or a rendezvous for one or more lynxes. Under the windfall were three well-worn beds and a well-worn trail leading to them. The beds consisted of a thin layer of loose moss scraped together. Near the den were a number of scats.

Don Herning told me that early in June, 1955, he had heard something crying along the road near his home a few miles from Fairbanks, and upon investigating he had found a young lynx with its eyes still closed. Possibly the mother had been carrying it and dropped it when his car approached. The baby lynx was raised and became very tame. Apparently it was born about the same time as the family I had observed.

During the mating season the lynxes do much yowling. In a letter dated March 6, 1956, Charles Ott wrote me that this year he had not yet heard the lynx yowl, but that the year before at "this time they were yowling regularly." One spring Ott saw a lynx walking slowly, "uttering a disconsolate yowl at every step," as though he were a rejected suitor.

The relation of the male to family life is not known. In winter, Ott has seen two adults and five young of the year together, but possibly the male joined the family after the young were moving about with the mother; or the adults may both have been females. Family groups, at least to the extent of mother and young, frequently remain together during the winter months.

Lynxes were occasionally seen traveling through the residential area (eight or nine houses) at headquarters. My brother, when he was talking to the superintendent one day in the administration building, saw a lynx pass the window. Ott wrote on March 6, 1956, that rabbits had decreased considerably, but:

The lynx are still seen frequently. They seem to come through the headquarters area here regularly. . . . A family group of three that have been seen several times still seem to be together. I saw them last week on Rock Creek. There was another family group of five that came through here regularly in the winter. I have not seen them for some time now, nor have I heard of anyone else seeing them. A few weeks ago Scottie and I spent nearly a whole day snowshoeing, looking for lynx, without seeing a fresh track. It was twenty below or thereabouts and our cameras did not want to operate properly. About three o'clock we went back to my place for a cup of coffee and had just settled down when I looked out the window. Here came three lynx around the end of the garage and alongside the house. Scottie and I like to broke a leg getting our cameras and running out of the house. The lynx went up the hill and into the willows back of the administration building, and we after them . . . no jackets, gloves, or caps. We never got a chance for even a long shot . . . we were icicles when we got back; the coffee really went "good" that time.

On March 20, 1957, Ott wrote me:

I just saw a lynx a short while ago as I came in from shutting down the pump. It was a very large, light-colored one. It appeared in fairly good condition and I swear the look in its eyes was actually friendly. It was sitting up under a tall spruce . . . we looked each other over for about five minutes. About seven magpies were in the tree just above the lynx and when he left they followed right after him. He crossed the road just at the curve and went down the bank to the creek and then down it. You could follow his movements by watching the magpies.

The daily life of the lynx, as we have noted, is hitched to the snowshoe rabbit. He has an unabridged predilection for them, and it is for them he is fitted for hunting. He often feeds on a mouse, a ptarmigan, or a ground squirrel in summer, but it seems that his mind is generally saturated with visions of Peter Lepus and that his imagination travels with difficulty beyond a rabbit dinner. This would all be very well, except that through the ages things have come to such a pass that he can't very well do without rabbits.

Since rabbits do not have planned parenthood, their fate is governed wholly by natural laws. For a period of years the rabbit statistician can confine his figuring chiefly to the process of multiplication. For the lynx, this makes for a rosy world. His food supply increases by leaps and bounds, and of course he also flourishes and has maximum family life. A day comes when there are hundreds of rabbits per square mile and a big lynx population. The woods are filled with life and activity. The big rabbit population, however, jeopardizes every rabbit, for now any disease can spread rapidly. And that is what generally happens in the far North unless food scarcity strikes first. A wholesale die-off of rabbits takes place. In a year or two the lynx finds the woods empty. Starvation is now in the land. The lynx stalks the woods, capturing what he can. Old hunting grounds are deserted, for where there were dozens of rabbits there may now be none. New habitats are explored and hunting of new prey takes place. In McKinley Park the lynx has turned to hunting sheep in the wake of a die-off. The lynx becomes gaunt and fails to reproduce and in a year or two is scarce or absent over wide areas.

Little by little, slowly, then more rapidly, the rabbits come back and again pyramid. The lynx recover somewhat too and become more plentiful. But the lynx numbers may in some rabbit cycles get off to such a poor start that they reach only moderate numbers before the rabbit crash again takes place. So there is some variation in the size of the lynx peaks and, perhaps to a lesser extent, in the size of the rabbit peaks. Such is the general pattern of the rabbit and lynx cycles.

The story of the lynx population in McKinley Park differs somewhat from that of most of interior Alaska, perhaps because it is on the edge of the rabbit range. Some rabbit peaks in interior Alaska by-pass most of the park and do not always synchronize with those in the park. Although the rabbits generally pyramid about every ten years, they did not pyramid in the park after 1927 until 1954.

A high rabbit population in the spring of 1955 decreased strikingly during the summer and continued to decline in 1956. Lynxes also declined, from high numbers in 1954 and 1955 to only scattered sign in 1959. But a few are still around, a nucleus for a future period of lynx abundance and prosperity.

3. The Ways of Grizzly Bears

THE NORTH SIDE of the Alaska Range is grizzly country. Old bruin may be found from the partially wooded terrain along the north boundary of the park to the glaciers at the heads of the many parallel river valleys. The entire country is his home, and one may meet bears anywhere from the river bars to the ridge tops. Regardless of where one camps, one is sure to have grizzly neighbors. Even in remote areas such as Mc-Gonnagal Pass they are ever on hand to utilize unguarded food caches. One group of climbers up toward Anderson Pass thought grizzlies "sure plentiful" there. When we were camping at the head of Savage River in 1922 and 1923, we saw bears so frequently that we assumed we were in the very choicest bear country.

In spring and fall one is especially likely to see bears on river bars digging roots, and they follow the same occupation on mountain slopes. In summer, high passes such as Sable Pass are especially frequented for grazing, and, in season, berry chomping may take the bears anywhere. Always they wander freely over their ranges, with few worries, taking care chiefly to avoid proximity to other bears.

The grizzlies vary in color from "straw color" to rich chocolate and black. A Sourdough told me one day that he had seen a mother with three cubs — "one lemon, one orange, and a chocolate!" Some of the faded straw-colored bears appear quite whitish when the sun is reflected at a certain angle, for a moment being taken sometimes for Dall sheep. The face is dark brown or blackish, and the feet and legs are blackish. In the fall of 1959 I saw an unusual cream-colored bear. The black legs and brown face contrasted sharply with the rest of the coat.

The head is large and broad. The facial profile is "dish-faced," that is, the forehead rises so abruptly that it is not in a straight line with the muzzle. The eyes are small and the nose somewhat squared off. The hump over the shoulder, along with the dish face, is useful in distinguishing the grizzly from the black bear. The claws on the front feet are long and but slightly curved, in contrast to the shorter and more tightly curved claws of the black bear. The long claw length shows up clearly in the track. Grizzly feet always have, relatively speaking, a dainty appearance, perhaps because they are covered with shorter hair than the legs. The hips are wide and solid; the stomach sometimes sags.

The weights of grizzlies have often been estimated, but few bears have been weighed. A bear a long distance from a scale always weighs most. One well-known bear hunter in Alaska, who had three dead bears in the hills at one time and was taking a naturalist out to see them, had all their weights estimated. But when he learned that the naturalist carried steelyards with him, the hunter began to hedge, and the closer they approached the bears, the lower fell the estimates. The actual weights of all three bears were far below even the reduced estimates. A very old black male, with much-worn teeth, that was shot as he was raiding a construction camp kitchen at Savage River weighed, after loss of blood, 650 pounds. Dressed weight (without head, hide, and innards) was 439. He was not very fat, having only a thin layer over his body. His live weight was probably close to seven hundred pounds. If he had been really fat, he might have weighed fifty to a hundred pounds more.

The mother generally nurses her cubs while lying on her back but occasionally nurses them from a sitting position.

I recall the first bear track I ever saw. It was my initial day afield in McKinley Park and my brother and I were crossing from Jenny Creek over a rise to Savage River, on our way to the head of the river. One lone track in a patch of mud is all we saw. But the track was a symbol, and more poetic than seeing the bear himself — a delicate and profound approach to the spirit of the Alaska Wilderness. A bear track at any time may create a stronger emotion than the old bear himself, for the imagination is brought into play. You examine the landscape sharply, expecting a bear on every slope as your quickened interest becomes eager and enterprising. The bear is somewhere, and may be anywhere. The country has come alive with a new, rich quality.

The track of the front foot is diagnostic because of the long claws, which leave their marks about two inches beyond the toe pad (this distance is about one inch in the black-bear track). The front-foot track frequently shows a small rounded depression back of the main pad mark; this is the impression made by the small posterior pad, which is so situated that it does not always touch the ground.

Although grizzlies generally travel about at random, they occasionally develop a characteristic trail. Short pieces of such trails have been noted in McKinley Park in woods bordering streams. I found a very deeply worn trail along Wood River in the mountains several miles east of the park. This definite trail was used by bears because of the lay of the land. The main channel of the river was flowing at the base of high, steep banks, and bordering the banks was a dense stand of spruce. Bears traveling up and down the river on the side I was on found it convenient to follow a trail through the woods. In using the trail the bears had characteristically stepped in the same tracks until two rows of deep track depressions had been worn into the hard ground. That the trail was much used was indicated not only by the track depressions but also by numerous rubbed "bear trees."

There is probably considerable individual and sexual variation in the home ranges of grizzlies, and females with cubs probably behave differently from those without. Some of the variations would be due to the lay of the land and the distribution and abundance of food plants. The range in spring, when roots are the chief food, is slightly different from what it is later in the summer, when the diet has shifted to grass or berries. But the changes due to diet would usually be rather local. My impression is that in the spring, when bears emerge from winter quarters, they travel more widely than later. A male probably wanders far during the mating period in search of a mateable female.

The females with cubs often confine their movements for most of the summer to an area less than a dozen miles in diameter. At Sable Pass we have seen females with cubs at short intervals for weeks in an area only six or seven miles across. Sometimes a bear will feed several days in an area and then climb over a high ridge to feed in another valley for a time. Thus, bears in Sable Pass or along Igloo Creek may move over into Big Creek, or they may go to the Teklanika River for a few days. Lone bears were known to occupy a limited area for a few days and then wander away, appearing and disappearing during the season. Their identification was uncertain, and data on them are not extensive. Several females that were in an area with spring or yearling cubs occupied the same area the following year. One female with three cubs was recognized in the Polychrome area during three successive summers; the third year she was followed closely by the two-year cubs.

The home ranges of bears overlap broadly, some bears having practically identical home ranges at least for most of the summer. Three mothers, each with yearlings, spent one summer on Sable Pass (a favorite grizzly area) and I once saw a fourth female with yearlings there; I saw a female with spring cubs a few times and another family of yearlings only three or four miles away. In addition to these, two large males and at least two breeding females were also in the area for several weeks. In 1950 two females, one with three cubs, the other with two cubs, spent the summer on Sable Pass. They kept apart as much as possible but often were near enough together to be acutely conscious of one another. Usually both females cooperated in keeping apart, like similar poles of a magnet. The presence of these two females on the same range was the cause of a tragedy, which I shall describe later. (In a garbage dump in Yellowstone Park I have seen thirty grizzlies wallowing together with bodies practically touching. Here, apparently, wild natural habits are being lost, and the dump is making of our lone philosopher bears a bunch of gregarious characters. They perhaps are gregarious, however, only at the dump. Along the Alaska coast the brown bears sometimes congregate loosely to feed on the migrating salmon, but are not congested as at a garbage dump.) A certain amount of hostility is perhaps one factor in keeping bears widely distributed over McKinley Park. I assume that a bear finding an area bearless or sparsely populated might tend to remain there, for they are "lone wolves" rather than gregarious under natural conditions.

Mating in McKinley Park takes place during May, June, and early July, when one may occasionally see in the distance a bear followed by a much larger bear with long legs and a stiff, long stride. It is the male following the female who at long last (unless this happens to be her first experience) is ready to breed again. Mating behavior seems to be quite varied, as the following observations indicate.

About May 10, 1941, I saw a female grizzly at Toklat River, near a small road camp. She spent most of the day digging roots and fed on garbage thrown away near the camp. About a week later she was joined by a young, lighter-colored male, slightly smaller. I saw them breeding on May 20. They were together most of the time and were often

playing together and hugging, much like the two bears I
have just spoken of. On May 22 it became a triangle affair,
for a huge dark male appeared on the scene. He endeavored
to drive the small male away, chasing him far up the moun-
tainside, back to the bar, and up the mountain a second time.
When the large male followed the female, she also ran away
from him. On May 23, at nine A.M., I saw the female digging
roots. A little later, a half mile away, the large male was following the small male across the broad river bar. The large
male grunted at intervals. The bears climbed up a slope,
where they fed on roots, the smaller one keeping an eye on
his rival. Later he descended to the bar and started across
to the female. But the watchful bigger male hurried down
the slope and on the bar came to within two hundred yards
of him. Both bears then galloped away, disappearing behind
a patch of woods, and then reappeared four or five hundred
feet up the slope on the other side of the bar. The small
male returned to the bar, started to cross it, but changed his
mind and came back to the female. The big bear, seeming to
be winded by the exertion, rested high on the slope for a
long time, then came down and was lost to view in the trees.
The female and small male wrestled, then fed close together
on roots. Later they wrestled again and the male grabbed
the female with his jaws back of an ear and tried to mount
her, but she rolled over. They continued to wrestle and play
for some time, then resumed feeding on roots. When I left
at two-thirty P.M., the pair was crossing the bar. I did not see
the big male after he came off the mountain and disappeared
in the woods. It appeared that the female was being true to
her first love and that the stronger bear was unable to win her
affections.

On May 24 the small male was in the woods. Later he was

joined by the female, which apparently had been chased, for not long after she arrived the large male came through the woods on her trail, grunting and bawling loudly at intervals. At his approach the other two bears ran away together. On June 2 I again saw the small male mating with the female.

The large male continued following the pair of bears at Toklat. On June 8, when I saw the three moving toward Mile Sixty-six, the large male was chasing the female. On June 9 I saw the three bears two miles from Mile Sixty-six, fourteen miles from Toklat. All three were sleeping only a few feet apart, on a point of rock. They lay sprawled on their backs, stomachs, and sides, occasionally changing positions or stretching a leg. Most of the time they lay on their sides. When I returned three hours later, they had moved to the gravel bar, where they were sleeping on their sides on the wet mud. The persistence of the large male had been rewarded to the extent of sharing the female. (In a report on the Kenya Parks of Africa it is said that two male lions shared a single female, behavior said to be unusual.) It was reported that the large male mated with the female on June 10. Soon after this date the mating activities of the bears apparently terminated.

These grizzlies had mated over a period of several weeks. The small male and the female were together for at least twenty-three days.

On June 12, 1953, I saw three bears on the north slope of Cathedral Mountain. A male was in possession of a female, and a bear slightly smaller than the male was hanging on the outskirts. This third bear appeared to be another male. I first noticed one of the bears at four P.M., as it lay on the slope, quite alert, for every few minutes it raised its head to look around. At eight P.M. I noted that three bears were in the area; earlier two of them had apparently been hidden by the alder and willow brush, which was quite tall. The large, dark male walked up to another bear, apparently a female, and they touched noses. Then the male covered her for about an hour. Much of the time the female wriggled about, apparently trying to escape while he held her with his paws in front of her hips and his head lying along her neck. Once she escaped but ran only ten or fifteen yards and stood, and the male again covered her. The third bear approached to within about twenty yards of the pair when the male first covered the female, but later went off a short distance, moving about a little as he watched the pair. The following two days the three bears remained in the area, much of the time hidden from view. On the fifteenth they were reported moving toward Igloo Creek.

It is interesting to watch a pair maneuvering on a slope. On June 3, 1953, I saw a large, dark male, with a huge, shaggy head, following a much smaller, straw-colored female. He kept herding her from below, as though his objective was to keep her up the slope. When she traveled, he traveled on a contour below her. Once, when a sharp ridge hid her from him, he galloped forward and upward to intercept her, but she seemed to have anticipated this and had doubled back. When he saw her again she was two or three hundred yards away. He galloped after her, his hoarse panting plainly audible half a mile away. But she made no real effort to escape, and he was soon herding her from below again. At noon they lay down about twenty yards apart, and a little later, when he went into a dip to feed, she hurried away as though she were playing a coy game with him. When he returned and found her missing, he galloped a short distance in two directions and, not seeing her, went back to where she had rested and from there followed her track. She was on a prominence, moving slowly, and after a gallop he was again directly below her. The chase continued all day. In the evening, as they were lying near each other, I saw the male get up and slowly approach her, but she quickly moved away.

Another pair I observed on Divide Mountain on June 30, 1947, went through a similar performance, the male herding the female from below. Once, when they stood facing each other about twenty yards apart, she raised her paws above her head and struck the ground stiffly, as bears do when bluffing. When she wandered among some large boulders, one of them was dislodged and would have struck the male as it bounced down the slope if he had not jumped aside. (I take this to be accidental.) The following day they were still in the same area, maneuvering as before.

A big shaggy male grizzly in the summer of 1959 kept company with two females. He was always recognizable by the limp in his front left foot. A limp is hard to analyze sometimes, but the left leg seemed stiff, raising his body a little as he walked. The left elbow seemed to protrude outward excessively. The long, unkempt hair over the eyes gave him an ancient, patriarchal look. I first saw him on June 13 as he was making his way with long, deliberate strides down a ridge in the Sable Pass area. He surprised a mother and two spring cubs that were foraging near a bank in a swale and galloped lumberingly toward them, but they hurriedly galloped up the slope away from him and did not stop until they topped the ridge and went out of sight. He continued to the main draw, where a miniature creek a few feet wide wound down through a narrow strip of tall willow brush. The huge fat hindquarters seemed to have an excessive sidewise wobble, and this, along with the hitch of the lame front leg, gave him a characteristic gait. I measured the tracks he left when he crossed a snow bank in the draw and found the hind track to be about ten and a half inches long. He was especially picturesque silhouetted against some of the snow banks he followed.

A week later, on June 20, I saw this male walking far behind a very blonde female. On the twenty-sixth, in the morning, he was following the blonde closely, and at times she was following him. That same evening this male was in pursuit, at a deliberate walk, of a very dark female that hopped along on three legs. She carried her right hind foot, on which she had a round bare spot just above the heel. I had first seen her in the area on June 17. He followed her to a point near the top of a ridge where at times he was only about fifty yards from her. They maneuvered about the ridge for some time before they both lay down. A little later the crippled female had moved far below, to the base of the ridge. The next day, June 27, the male was breeding with the blonde when I discovered them. He held her for about ten minutes. When she escaped, she sat down a few yards away. The crippled female was at the time resting about twenty yards away; soon she moved over a ridge out of sight. The male followed and the blonde brought up the rear. The male continued following the crippled female for about fifteen minutes and she kept moving well ahead of him. Later in the day I saw this same procession.

On the twenty-eighth, the three bears were together and the male was breeding with the crippled female. Earlier in the day the crippled female had been very coy and at one time was half a mile from the other two bears; I saw her lying on her back, kicking all four legs in the air. Twice, when the blonde was only a few yards from the male, he

Mother with spring cub in early summer.

Mother with yearlings in early summer.

made a few bluffing jumps toward her, hitting the ground hard with his front paws. But both times she returned at once to him after retreating a short distance.

This miscellaneous maneuvering continued for the following few days, the male following one or the other of the females. On July 1, the male followed the blonde part of the time but paid more attention to the crippled female. Her foot, by the way, was much improved; although she still carried it when loping, she used it in walking. The blonde, after July 1, wandered farther away from the other two bears, but she was near them on July 4 when I again saw the other two breeding.

On July 8, when I discovered the two crippled bears about ten o'clock in the evening, he was holding her as she squirmed. After five minutes she broke away, moved off twenty-five yards, and waited while the male overtook her again. This time he held her for twenty-five minutes and for much of the time she continued squirming in his grasp. When I left they were resting about five yards apart.

The last day I saw the crippled pair together was July 10. The blonde had left a few days earlier. I continued to see both crippled bears in the Sable Pass area for a few weeks, the male until July 26 and the female until August 4. (They probably moved away in search of berries, the crop on Sable Pass being below par.) The blonde female, I did not definitely recognize during this period.

I saw a second large male in the area during the mating period, on June 12 and 28 and July 4, but he was not near the mating bears, usually being a mile or more from them. During the mating period, the females paid no attention to each other. Each one had been in the company of the male for at least two weeks, but I observed the blonde with the male about a week before the cripple was with him; apparently the blonde left him about a week before the cripple departed.

Observations on nursing yearlings and two-year cubs are significant in regard to the breeding interval of female grizzlies in the park.

Mother with two-year cub (third summer abroad).

In 1953 I was surprised to see a yearling cub nursing. On a few occasions in the literature a yearling black bear and brown bear cub had been reported nursing, but it was assumed that these instances were exceptions. Since my first observation of the nursing yearling, I have had opportunities to check on thirteen mothers that were followed by yearlings. In every case these one-and-a-half-year cubs were nursing; the interval and duration were similar to what I had observed in spring cubs.

My observations suggest that cubs also regularly nurse the early part of their third summer abroad. In the seven families I have been able to check in this regard the females nursed their large 2½-year cubs. The latest nursing noted was July 12. All observations indicate that families generally break up during the third summer between late May (one record) and September.

The observations of nursing yearling and two-year cubs

suggests a minimum breeding interval of females with cubs of at least three years. In some cases in which I have observed females followed by two-year cubs throughout the breeding period, the interval appears to have been at least four years. It is also significant that, of the two dozen or more mated females I have observed, none were followed by a yearling, and only one by a cub older than a yearling.

The one, two, or three cubs (the usual numbers) are born in midwinter in the hibernation den. They are extremely small, weighing only about one and a half pounds. They still seem very tiny when one sees them abroad with the mother in early spring.

Grizzlies in McKinley Park go into hibernation in October and emerge in early April. In 1939 I saw bear tracks on April 17, the first day I was in the field; and in 1941 I saw the first bear track on April 8.

On October 11, 1939, when there was a foot of snow on the ground, I saw a grizzly digging a den on a steep slope far up a mountain. In digging, the bear disappeared into the hole, then came out tail first, pawing the dirt out of the entrance. At intervals he pawed back the pile of dirt at the entrance, and dirt rolled far down the slope over the snow. The entrance was just large enough to permit the bear to enter. The following spring I saw fresh tracks leading away from the den; undoubtedly it was used by the bear for his winter sleep. When I climbed to the den later in the summer the chamber had caved in, so it could not be used a second year. The chamber was four feet from the entrance and was about five feet in diameter. The floor of the burrow led upward at an angle of about ten degrees. Where the den was dug, the mountain sloped at a forty-five-degree angle. Other dens, also caved in, were in similar situations.

A bear may have a den in mind long before denning time. On July 22, 1953, I found a freshly dug den on a slope. The bear had recently dug a tunnel with a slightly upward slant, twelve feet long, just under the sod. The entrance was about twenty-four inches wide and twenty-seven inches high. A chamber at the far end had not yet been dug. When a companion and I visited the den again on August 23, we learned that the bear had been back and dug a chamber at the far end that measured four by three feet, the longer dimension being at right angles to the burrow. The chamber may not have been finished at this time. A quarter of a mile from this den was the fresh beginning of another burrow that was only about five feet long.

Six years later I again visited this den and found it still usable. The sod roof was especially firm and no doubt accounted for the long preservation of the den. In front of the entrance a number of cinquefoil bushes had been nipped off in past years and brought into the chamber for bedding. Also in the chamber were remnants of dry grass and herbaceous material. Apparently the den had been used at least once, and possibly oftener, since it was dug in 1953.

On another slope, quite steep, I found an old den whose roof had caved in some years ago. The broad mound of dirt at the den entrance was now covered solidly with herbaceous dogwood. When I visited this den in 1959, I learned that a bear had more recently started a den just below this old one. It was unfinished, but extended six feet, and part of this distance was under the dirt mound of the old den. Had the former occupant of the old den returned some years later? Or perhaps another bear recognized the old den as an old den and it had suggested den digging to him.

Bears do not go into a comatose state during hibernation

as do the ground squirrels, but can be activated readily at all times. They emerge from the den fat but are said to lose weight later. Food resources may be limited when they first emerge, so it seems logical that they should lose some weight then. Bears seem to be traveling a good deal at this time, a way of life which also would be conducive to loss of weight in the spring.

4. Bears and Squirrels

THE GRIZZLY IS, for the most part, a vegetarian, but perhaps largely by necessity rather than by choice. During the span of about seven months that the grizzlies are active in the park, their feeding habits pass through three marked phases. In the spring, up to early June, the chief food consists of roots. During June and July they feed mainly on green vegetation, chiefly grass and horsetail. In late July, when berries become available, and during August, September, and October, the bulk of the food consists of berries. Roots also become part of the late fall diet. These principal foods are supplemented by others. A few ground squirrels are caught and eaten at all seasons, and carrion is always highly acceptable.

Throughout May and into the first week of June the majority of the bears I saw were digging for roots. The root chiefly sought is a pea vine (*Hedysarum alpinum americanum*) that grows abundantly on old river bars and on many mountain slopes. Generally they seek the roots on the vegetation-covered bars or low slopes, but often they seek them far up on the mountain tops. Some of the favored spots, covering a few acres, are dug so extensively that they suggest a plowed field. In some of these places I saw bears digging for roots day after day. I was interested in finding much sod turned over by bears just above some cliffs where the sod terminated. Apparently it was easy to turn over the sod here, where it lay like a carpet with an edge exposed. The bears' activity here had increased the erosion process.

Although roots are eaten mainly in the spring, I also found a number of places where they had been dug during September. Feeding on roots is probably resumed at this time because some of the other foods are either less available or not palatable — also, the roots may become more palatable than earlier.

To get at the roots, the bear usually places both paws on the ground and thrusts back with the body until a chunk of sod is loosened. This is turned over, the free roots are devoured, and then, with a paw working slowly and lightly, more of the tender roots are uncovered and eaten. I have watched bears feeding in this manner for several hours at a time. The fleshy roots consumed range up to a half inch or more in diameter and resemble dandelion roots.

In June and July the grizzlies may be seen chomping away at grass, horsetail (*Equisetum arvense*), and herbs such as sourdock (*Oxyria dygna*) and *Boykinia richardsonii*. I have had some difficulty getting their favorite grass identified. It is a juicy-stemmed species that grows in moist areas. According to the latest identification I have obtained, the scientific name of this grass is *Arctagrostis latifolia*. The bears know it well.

Blueberries, cranberries, and buffalo berries are grazed with great vigor. Berries, leaves, and twigs are all gobbled up together. The bears have too much eating to do to be finicky and selective in their berry eating. Even with their rough-shod methods they are kept rather busy filling their paunches; furthermore, the food passes through them rapidly, judging from the slight digestion it seems to receive and the frequency with which scats are deposited. On one occasion I saw a bear leave four scats behind her during a single hour.

On August 1, 1940, I spent part of the day watching a mother and her cub feeding on blueberries. She fed continuously and hungrily from nine A.M., when I first saw her, until eleven A.M. She lay down for half an hour, then fed steadily from eleven-thirty A.M. until four P.M. She lay down for twenty minutes and then commenced to feed. But in half an hour she lay down again and I left her. The spring cub picked at the berries only part of the time and rested frequently while the mother fed.

Ground squirrels are eaten at all times. This is the one animal, aside from mice, which the bear can hunt methodically with some success. Spring, summer, and fall diets are all supplemented with ground squirrels, but usually in small amounts.

Much time and energy are generally used in seeking squirrels. Sometimes luck is with the bear, and ground squirrels may be captured in five or six successive excavations, but more frequently a bear may dig out or prospect two or three holes without success.

On August 8, 1939, I came upon a mother bear who had dug a trench fifteen feet long and was still digging near the middle of it. Her three cubs were digging in various parts of the trench, perhaps in imitation. The mother would place both front paws on the sod and push downward and pull back until the piece gave way. Sometimes the bear appeared to push down on the sod to scare the squirrel. (Several times I have seen a squirrel emerge from a hole after the bear has jarred the sod with its paws.) She would paw out the dirt and take frequent sniffs at the mouth of the hole where she was working. Once she reached into the hole with her arm, lying on her side so she could reach in farther. At this point the cubs all crowded close in expectation. The mother then moved to another entrance above the place where she had worked, and as she was pressing down the sod with her front paws to loosen it, the ground squirrel scurried forth from the hole. The bear jumped for it, but it dodged to one side and managed to escape from four pounces, the last time leaving the bear flatfooted off to one side. But with the next pounce the bear captured the squirrel. The cubs sat close by, watching the mother chew the squirrel, but she made no offer to share it with them. One cub did get a morsel which dropped from her jaws. More than half an hour was expended in catching this squirrel.

On July 21, 1940, a grizzly discovered a squirrel under a large snowdrift perhaps twenty-five feet across. There was an entrance on the upper edge of the drift and one on the lower side. The bear kept going back and forth from one entrance to the other. In going from the upper hole to the lower one, he often slid down the drift. He would poke his head into the upper entrance to sniff the squirrel, then return to the lower hole to continue digging into the snow. He dug into the drift so far that he was completely hidden. Once while the bear was in the lower hole digging, the squirrel came out through the upper hole, sat up straight to look around, then scurried away from the drift. The bear made a few more trips up and down the drift after this, then lost interest, seeming to realize that the squirrel was no longer there. He stood on the drift hesitatingly, seeming to be deeply disappointed. A whole half hour of strenuous work, all for naught! He walked off to a patch of grass and grazed, a meal perhaps not as tasty but one more certain.

On October 9, 1940, I saw a female bear wandering over a snowy slope and examining half a dozen ground-squirrel burrows with her nose. Finally she poked her nose through the snow into the entrance of a burrow and sniffed hard. One could tell by her actions that the scent was hot. She tore away some sod and dug down through the dirt. It took her only a few minutes to reach the squirrel nest, out of which she removed a ground squirrel already in hibernation twenty inches below the surface. Only the upper two inches of the ground were frozen at the time. While the mother chewed, her cub sniffed about and did some digging of its own in the loose dirt pawed up by the mother. The mother examined another squirrel hole, dug into it a little, then wandered off to feed on blueberries and cranberries.

On August 4, 1941, a lean bear hurried up the slope of a basin, turned, and followed a contour line in search of ground squirrels. It investigated a number of burrows with its nose as it hurried along, intent, it seemed, on finding a squirrel at home. Once it broke into a gallop but stopped abruptly at some holes and began to dig. At first it dug with one paw, at the same time keeping a watch on other holes from which the squirrel might emerge. Later, the bear used both forepaws in digging but stopped at intervals to look around for the squirrel. After much digging with no success it moved on, again breaking into a gallop. It stopped at another burrow, where it made a large excavation, still with no success. A magpie sat on a willow nearby, watching it dig and waiting for scraps. The bear wandered away from this failure and fed on blueberries, which on this particular slope were not abundant that year. This bear had spent about forty-five minutes seeking squirrels, with no success. It could not afford to gamble further on the ground-squirrel hunting and returned to the staple diet. After filling up on berries the bear would probably again try its luck and dig after ground squirrels, for it seemed hungry for a taste of meat.

If there were not considerable uncertainty in ground-squirrel hunting, the bears would no doubt devote more time to it. But there is a limit to the amount of gambling they can indulge in, and even if successful they must return to foods which are available in quantity to fill their rapidly emptying paunches. The ground squirrel has been referred to as the staff of life of the grizzly, but it is only a side dish.

The grizzlies are much interested in the calves of caribou, moose, and sheep, but in the park they get relatively few, especially of the latter two species. There is not space to discuss bear incidents pertaining to all these species, but I shall relate an incident concerning caribou that is no doubt repeated several times each year. Because caribou are abundant, a few calves are captured while they are still very young. This predation is chiefly by bears that are living where the calving is taking place.

One evening our attention was attracted by a band of caribou galloping into view from behind a ridge. Their rapid gait suggested that they were running from something. Surely the enemy would be right on their heels! But an interval passed, and when it seemed that no pursuer was going to appear, a grizzly galloped into view, as though he were still an important part of the show. He charged the caribou, which were now standing in a scattered band watching his approach. He galloped with energy and concentration. When he crossed a soft spring snowdrift, he sank deeply; undeterred, he plowed through by the sheer surplus power in his legs. When he neared the caribou, they sped quickly away and scattered. But the bear persisted, chasing one group after another. Once, as he came down a slope toward a band, he seemed to put on his last ounce of speed for a final effort, but to no avail. For fifty minutes he galloped hard after the caribou before he gave up and moved into a draw, where he tarried, either feeding or resting, for he did not emerge while we watched.

With so little opportunity of catching any of the caribou, one is led to wonder that the grizzly should expend so much energy in futile chasing. The answer is that his persistence had been rewarded earlier in the season, when he used the same technique and was able to capture calves too young to escape. On the day we watched him, the calves were no longer vulnerable. The bear did not realize that his timing was off, that the calf-hunting season had passed. I have observed many similar out-of-season chases.

When a bear has feasted on a large carcass, he often covers it by scraping over it vegetation, sod, and debris from the surface a dozen or more feet around. One grizzly had covered the remains of a sheep carcass and continued from time to time to rake the sedges and some sod from the surrounding area as far as twenty feet away. Sometimes, before covering the carcass, he drags it a short distance. The bear often leaves the cache after feasting, or he may lie down on it and drive away meat-hungry bears or other carnivores that come along.

Occasionally one sees a bear with a few porcupine quills in its nose, and one animal was reported stuck up in the face very generously. One day I watched an old bear, which was feeding in a swale, stop feeding and watch a porcupine that came waddling over a knoll only a few steps away. This bear knew porcupines and permitted it to detour to one side. In the spring of 1959 I saw a young bear, three or four years old, limping on a front foot as he approached Sanctuary River, which he crossed in spite of the fast current. Two days later this bear was shot at Savage River camp ground, six or seven miles away, because he was disturbing the campers. Three of us examined the injured foot and found the remains of old porcupine quills, and some festering in the foot and under the shoulder blade. The claws on the crippled foot were excessively long for lack of use, while those on the other front foot were shorter than usual, probably because all the root digging had to be done with this one paw. In the same year I saw a two-year-old cub carrying one of its front paws and chomping its jaws as though it had quills in its mouth. A yearling cub was also carrying a front foot. It seems likely, in view of the quills found in the dead bear's paw and in the face of some adults, that these two cubs had pawed at a porcupine. The two-year cub almost certainly had encountered quills.

Naturalists have conjectured much concerning the purpose of bear trees. Ernest Thompson Seton wondered if they were not registers of some sort, a place to let the bear world know who had been that way and how big he was, as indicated by how high on the tree he could make his mark. The trees no doubt do serve as a register, because a bear using one would get the scent of the last recent visitor.

From my experience, however, such use would be inadvertent. I have watched many bears of all sizes and both sexes using bear trees or their equivalent, and always they used them as something to rub against, either their sides, stomach, rear, or back, especially the back. A bear tree will always show bite and claw marks, but these appear to be a secondary activity in connection with the rubbing. The claw marks are sometimes made when the bear is standing on his

(*Top*) The author climbing Eagle Summit at the end of an all-winter trip.

(*Bottom*) En route to Brooks Range. The lead dog is Irish, a 140-pound pup; the other two are quarter-bred wolves. *Photos by O. J. Murie*.

(*Top*) An adult lynx and four young during the abundance phase of the lynx cycle. *Photo by Charles J. Ott.*

(*Bottom*) Gail and Jan Murie crossing a glacier stream in McKinley Park. *Photo by the author.*

(*Top*) Lynx in midwinter. Note long ear tufts and whitish throat ruff bordered with black. *Photo by Charles J. Ott.*

(*Bottom*) Lynx walking; note long legs and black-tipped tail. *Photo by the author.*

Grizzly bear contemplating a ground squirrel's burrow. *Photo by Charles J. Ott.*

(*Top*) Yearling cubs with mother in early summer.

(*Bottom*) The same trio one year later. On this day (May 18) the 2½-year-old cubs nursed four times. Two days later they had separated from their mother. *Photos by the author.*

(*Top*) Fox cubs, McKinley Park.

(*Bottom*) The mother fox, split-ear, yawning. *Photos by the author.*

(*Top*) Old man wolverine. *Photo by the author.*

(*Bottom*) The coyote was chiefly to be found in the lowland, snowshoe rabbit country. It was so scarce in the Park that it had little impact. *Photo by Charles J. Ott.*

hind legs and holding the trunk with his forepaws. One grizzly, as he stood facing a pole, reached up with both paws as high as he could, as if in a stretch, and left a few marks up high. A black bear who had rubbed his back against a tree rubbed the ear region next, but to get greater pressure he raised a forepaw above his head and back of the tree and thus held his head pressed close to the tree while he rubbed. He did this with the other forepaw too, when rubbing the other ear, and repeated the whole operation on another tree. The paws left marks on the bark.

The bears seem to bite into the trees idly, as a dog will bite at a stick. One pole that had been used a great deal by the bears had a deep bite notch four feet from the ground, and another about five and a half feet up. Possibly one mark was made when a bear was sitting against the pole and another when he was standing erect. In these two positions he turned his head and bit into the pole. I once noticed that a bear that had rubbed its side against a bridge rail had bitten out two large slivers from the railing.

Most of the bear's time at a bear tree is spent standing on his hind legs rubbing his back. This involves all kinds of contortions. Often the bear rubs his back by raising and lowering himself, as if chinning himself, except it is the hind legs he is using rather than the arms. In this operation, the tail region alternately is touched to the ground and raised a foot or more. Sometimes more efficiency in rubbing is obtained by wriggling back and forth as he moves downward along the pole. The posterior regions are rubbed when he is standing on four feet, facing away from the pole; the sides may be rubbed while standing on two hind feet or on all four feet.

At the base of a well-used bear tree in the open, a circular patch on the ground is worn smooth. The bark also shows wear from the rubbing, and hairs cling to the pitch or are wedged into crevices of the bark. The most-used bear trees are located in strategic spots where bears frequently pass. In a woods where trails are used, many trees along the way will show much use as will trees convenient to river bars, which bears are wont to use for highways. A lone tree, or the last one at the termination of a grove, is often used by the bear because of its strategic location.

Above timberline a willow clump, rock, cut bank or even the tundra may serve as a bear tree. Once I watched a mother bear trying to rub her rear while sitting in the tundra, but there was no fulcrum, for she held her hind feet in the air as she sat. She performed many ludicrous contortions and a great swinging of arms and shoulders, with only a modicum of success. After scratching her chest lightly with a paw, she lay down, legs up, to scratch her back, but again lacked a fulcrum. Such strenuous wriggling, jerking, and futile waving of four legs in the air was never surpassed.

A pole or log found lying in the open is usually an occasion for a rubbing and a body massage as the bear rolls and pushes over it. But a genuine bear tree is preferred to these make-shifts.

Bears have an uncommon predilection for human foods, whether in the form of garbage, or groceries in a cabin. Accessible garbage is the chief cause of bear trouble. First it attracts bears, then it continues to hold them in an area so that they become unafraid and are soon breaking into tents, trailers, or cabins in search of more food. Human contacts follow, and incidents occur in which people are harmed, sometimes seriously. The bears become pests, no longer interesting wild creatures with natural habits. The usual ending to the story is damage of property, injury to humans, and death to the bear.

When camping in bear country I have always burned all garbage and all cans that have contained food, to destroy the food odors. With these precautions, I have had very little bear trouble. It may be impractical for large camps and hotels to dispose of the garbage by burning, and it then becomes necessary to haul it out to a garbage dump. But it would seem practical to surround such a dump with a bearproof fence. This is generally considered too expensive, but it is chiefly a matter of what we think important. A rather large road camp took the trouble to burn its garbage regularly, and as a result the camp had very little bear trouble.

If food is stored in cabins, strong bearproof shutters should be used to protect the doors and windows. As a partial alternative, the food could be placed in a cache on top of four poles. I believe it would be desirable to build a picturesque cache at each of the outlying cabins in the park and store provisions in them instead of in the cabins.

In some areas, bear trouble has been reduced by livetrapping the bears and transporting them to distant areas, away from habitations. To minimize bear trouble, a combination of all precautions and remedies is needed. In national parks it is desirable not to have any garbage available to bears, otherwise they will be attracted to habitations and eat garbage and not be living their normal, primitive way.

For some reason, the public is unafraid of bears. Perhaps this is because real bears so closely resemble the Teddy bear. This attitude is justified to a certain extent because bears are, on the whole, rather good-tempered and well-behaved. But the danger lies in their potentiality for causing serious injuries, and the uncertainty of their behavior. A half-hearted attack or the mere casual swipe of a paw can cause a damaging or fatal wound.

When a friend of mine, about to embark into bear country, inquired about the danger of bears, I replied that he had nothing to worry about, that he could travel the wilderness with a light spirit, that all he needed was faith. The chief difficulty, I pointed out, was to preserve one's faith, that as one gains bear experience one tends to lose faith, but still if the faith is kept all would be well. The wariest people in the hills are the trappers and bear hunters; yet, after all, they prefer wandering over the hills to crossing streets in modern traffic. The moral is to respect the bear's potential for causing injury and to keep at a respectful distance.

Can the grizzly be saved for the future? It disappeared in California and in many other states. Will it be with us out of mere chance, and disappear when some strong dollar value takes over bear country? Multiple use has frequently become a shibboleth in public-land policy, but not all uses in a single area are always compatible. I saw grizzlies disappear from a public range when cattle were introduced — the two species were not compatible. The interests of 150 million people were sacrificed for those of three or four. On the other hand, wilderness recreation and grizzlies are compatible. Alaska is a big country, but now is the time to recognize and preserve those intangible values in our fauna and flora so important to our culture.

5. Of Bears and Men

IN DISCUSSING the ferocity of the polar bear, Ernest Thompson Seton wrote: "Before me is a pile of data dealing with the moods and temper of the Bear. One portion proves that the creature is timid, flying always from man, shunning an encounter with him at any price. The other maintains that the White Bear fears nothing in the North, knowing that he is king; and is just as ready to enter a camp of Eskimo, or a ship of white men, as to attack a crippled Seal."

My own experiences deal mainly with the grizzly bear, and I also have two piles of data. One pile is much bigger than the other, perhaps a hundred times, and it proves that grizzlies are timid and run away when they see a person. There is not much to tell about a bear that runs away, so I shall tell mostly about bears that did not run away — at least not at once. That, of course, is hardly fair to the bears, because it is likely to give too strong an impression of the ferocity of bears. They are not out looking for trouble. But, then, what I have to offer may contribute to an understanding of bear nature and be to his benefit in the long run.

Good bear data are not easy to gather. There are, of course, hundreds of bear stories — but a bear story is like a fish story: the teller feels he has considerable latitude so far as accuracy is concerned. He eyes his audience, and if they show the least bit of detachment, the figures for distance, size, and weight are generously manipulated, the adjectives strengthened, and even slight additions made to the plot. The story must be good. As it travels the rounds, it changes color and shape, always for the better. Then too, it is not easy for anyone concerned in a bear incident to know just what did take place. Too many things are going on at the same time, and what was happening gets mixed with what one is thinking. Sometimes the account must be changed to conform to the game laws. Another angle is brought out by Tolstoy in telling about a hero of the Napoleonic wars: "He [the hero] described to them his battle at Schöngraben exactly as men who have taken part in battles always do describe them, that is, as they would have liked them to be, as they have heard them described by others, and as sounds well, but not in the least as it really had been. Rostov [the hero] was a truthful young man; he would not have intentionally told a lie. He began with the intention of telling everything precisely as it had happened, but imperceptibly, unconsciously, and inevitably he passed into falsehood."

Bear stories have certain characteristics. Distances have a way of shriveling. A bear one hundred yards away will, in the telling, be only fifty yards or only fifteen feet away. The reduction consists of an honest underestimate of the distance due to the emotional impact of the bear and the desire to increase the excitement of the story.

Then it occurs to me that no one ever sees a small bear. All the story bears are big! And I have observed a keen ability to estimate accurately the height and the weight of a bear. This can be accomplished at surprising distances.

I do know, however, of one man who failed utterly to measure a bear with his eye. It was on Jarvis Creek in the

Alaska Range, and the hunter was after specimens of an odd, light-colored grizzly for a museum. From near the top of a mountain that he had climbed, he saw far below him on a gravel bar an animal, moving slowly. He could see the powerful forearm reaching forward as he walked — a big grizzly. A stalk was begun. Halfway down the mountain he stopped for another look. Through the field glasses he saw a small, roving band of caribou traveling up the gravel bar toward the bear. He sat and watched the caribou draw nearer. Here seemed an opportunity to observe something dramatic; perhaps the bear would try to capture a caribou. Closer and closer came the caribou, walking along with heads low. Then the caribou in the lead was passing only a few feet from the bear — it appeared to be no bigger than a porcupine — then it dawned on him that his bear *was* a porcupine!

The hunter was my brother, and he was quite ashamed of his error in judgment, and for a time said little about it. Subsequently, an old-timer told him about a similar experience. And one day, up the Robertson River in the Alaska Range, he and his companion, Tom Yeigh, an outstanding woodsman, caught a momentary glimpse of an animal disappearing in the brush. Yeigh thought it a porcupine — it was a big grizzly! Even after hearing about these incidents, I once watched a grizzly from a mountain top; he was near the base, and I watched him for some time before I realized with chagrin that I was looking at a porcupine. Long ago I learned that size is deceiving, but I go on talking about big bears, now and then becoming scientific enough to say, "It appeared to be a big one."

A more effective way of measuring bear size than calculating with the eye is to measure a bear track. One must assume that all bears have the same relation of feet to the size of the body. That is, a big bear has big feet and a small one has small feet. One does not need a tape measure, but one must know the size of one's boots, for that is the unit of measure. It goes this way: "Now, I wear number twelve shoe pacs. I put both feet, side to side, in the track of that bear, and there was room to spare"— with variations.

Of course, it doesn't matter how big a bear is, if the point to be made concerns the dangerousness of the animal. A small bear can be as dangerous as any big bear if she happens to be on edge and a little cranky. If cranky, it may be because you have come entirely too close, or because she is at a carcass, or because she has cubs, or is wounded. A wounded bear is by far the worst. All bears beyond cubhood are powerful enough to overcome a person without half trying.

I gained a special respect for the power of grizzlies during two summers in Wyoming, where cattle had been placed on grizzly range. Some of the grizzlies were occasionally killing some of the Herefords that had invaded their domain. One day, about noon, I came upon a turmoil of over a hundred cattle, bellowing and bawling. Lost calves were searching and bawling for their mothers, and mothers were trumpeting hoarsely for their lost calves. All the animals were on the move among the willows on a broad, grassy bottom land bordering a stream. Something unusual had happened to cause this major disturbance. A grizzly was undoubtedly on the premises, probably feeding on a victim.

I moved cautiously toward the general area from which the cattle were moving. A raven flew by with a scarlet piece of flesh in his bill, and I projected his flight back to locate the scene of the tragedy and moved toward it more cautiously. My approach had probably been noted, for the bear had left, but his two victims lay where he had felled them, still hot.

One animal was a large yearling, weighing perhaps four or five hundred pounds, the other was a spring calf. The bear had evidently grabbed the yearling in his great arms and had bitten it in the base of the neck, crunching the vertebrae. He had also bitten through the vertebrae of the "small of the back," corresponding to one's belt line. It needed a powerful bite to penetrate a vertebra protected by thick layers of muscle and hide.

I once saw a group of Civilian Conservation Corps boys creep up on a mature grizzly to get snapshots. The boys did not, of course, realize any risk. Maybe the bear was overwhelmed by their numbers, for he retreated, but any prospector would rightly denounce such rash action. In Yellowstone Park the tourists have amazing confidence in the benevolence of the black bears, and the majority of them experience no ill effects from their confidence. But each year many indiscreet people are scratched and bitten, sometimes very seriously. One can jaywalk in traffic and perhaps escape for a long time, but a certain percentage will fall victim.

Some will take no chances. A section man on the Alaska Railroad recently told me about a small surveying crew that had a job half a mile up the track. The section man offered to take them on his gasoline speeder, but they said that walking was good. In a short time the crew was back, saying that they would take that ride. They had seen a fresh bear track in the new snow and concluded a bear was in the neighborhood. The tracks had scared them, and they didn't care who knew it. There is nothing like a bear track to arouse the imagination.

Some years ago in the Alaska Range we were building a corral and lead fences across a narrow valley to catch live caribou as they came over a pass. They were to be used for cross-

ing with reindeer to improve reindeer herds. Our camp was a
mile from the corral, and we made the round trip twice a
day. No one worried much about the occasional grizzly we
saw on the hillsides except one of the experienced bear hunters in the group, who invariably carried his rifle. He said he
had had some close calls.

But I know a nature photographer who is fully confident
that bears will not harm him, and he is still alive. I know a
naturalist who usually carries a gun in bear country but is
casual about it and often travels unarmed. One could argue
strongly for being safer without than with a gun — it is the
wounded bear that is the greatest danger.

Perhaps the best way to tell about the ways of bears is to
illustrate with a few incidents. But may I point out that these
few became newsworthy and that countless times the bear
hurried away to the hills when he saw us.

In the summer of 1944 I arrived in Jackson Hole, Wyoming, in the midst of a nationally publicized bear episode.
The hero, Farney Coe, was caretaker on Cissy Patterson's
dude ranch. So far as I know, Farney told a straight story, but
he never told the climax as it was published. Farney said that
while he was out on a beaver dam he had been attacked suddenly by the mother of two cubs. With hoarse grunts the
mother had sunk heavy canine teeth into his arm and shoulder. Farney lay limp on the dam, playing dead, and the
mother bear left him, satisfied that she had completely van-
quished an enemy and saved her cubs. Farney said he gave
her time enough to be gone before he stirred. But the bear
was still on the shore, and when she saw Farney rise she again
charged. This time Farney changed his tactics; he picked up
a stout aspen limb which the beavers had used for building

the dam and swung it hard and true and, according to the stories, killed the bear.

Authorities were questioned about the probability of a man killing a bear with a club. The director of the Washington Zoo, experienced with animals of all sizes, doubted the likelihood of such a killing, according to one news report.

Half a dozen versions of the story were circulating, and everyone who told it differed on important details. I saw Farney one day in front of the Crabtree Hotel, where he was convalescing a day or two, and he obligingly showed me the tooth wounds in his arm and told me the story. I asked him if he had killed the bear. "No," he said, "but I 'it 'im so 'ard I guess she went off and died in the brush."

The story that he killed the bear was still circulating five years later. In a weekly magazine for November, 1949, the story about the killing was again published. It said, ". . . The animal fell, stunned. Coe, pressing his advantage in the primitive struggle, rained blow after blow until the shaggy beast lay lifeless."

A few years later, I learned of a rancher who had been wounded by a grizzly. I went to see him and found him in his barn trimming the mane on one of his fine-looking horses. "Getting ready for haying," he said. He wore a black, silky, full beard, talked in a low, quiet, modulated voice, and was a natural-born storyteller.

He told me about the female grizzly. He had gone out to earmark a newborn calf, because if he didn't his "cow would be coming home with someone else's calf!" He said he had left his saddle horse on one side of a brook and looked around for a shed-moose antler he had seen one day earlier that spring. In jumping across the brook, he landed beside a bear cub, which bawled loudly. Tom knew that the mother grizzly would be nearby, but before he had time to move she was upon him. He threw his arm across his throat and half turned. The bear struck him a glancing blow on his chest that stunned him and sent him whirling. (He showed me the rips in the leather vest.) As he pitched into the brush he felt a jerk or yank on his foot. By this time the cub, which had run in the general direction of the horse, was being chased by the rancher's dog and was again bawling. Then the mother left the rancher as though he were "something hot" and chased after the dog and cub. He lay there stunned for some time but managed to get home.

He showed me the rubber boot he had been wearing when the bear struck him as he pitched from the first blow on the chest. Each of the five claws had ripped through his rubber pack for a distance of about five inches, and part of the boot was literally in ribbons. The rip made by the middle claw was bloody, this claw having made a nasty wound in the foot which took a long time to heal.

The incident aroused no hatred in his heart. He said that any animal in like circumstances would have attacked. A few days before, he said, a badger with two young had made him climb a fence. I heard other versions of the bear story; there was no doubt about a bear having made the attack.

One of my own experiences with grizzlies took place above the timberline while I was climbing Sable Mountain in Mc-Kinley Park. Bears were commonly seen in this area, and I was not surprised to see a grizzly off on the side of the moun-tain following, in general, a contour line and coming my way. The thing to do was to get out of the bear's path, especially in this area where the bears were accustomed to seeing people and were rather unafraid. No one in the hills relishes a fearless bear. I should have done a little differently from the way I did; but, wanting to save time and walking, I figured out that I could pass the bear without his seeing me.

Between us was a good-sized hump on the slope. The bear, judging from his line of march and the lay of the land, would pass *above* the hump, and there would be no traffic jam if I walked *below* the hump. But a bear's actions cannot be calculated that closely — he starts off in one direction as though he were going that way for at least ten miles, and the next time you look at him he has reversed himself.

I followed the low trail, and as I was coming under the hump, there, a little way ahead of me, was the bear, also fol-

My bear rose up on his hind legs and he looked big.

lowing the low trail. Above us was a precipitous slope; below, rough, steep cliffs — it was a one-way thoroughfare. I stopped and the bear stopped. There is a story in our family that a Scandinavian grandfather faced a bear for hours with only a knife in his hands, and the bear was finally outbraved. But this thought was canceled by the story of a bear-hunter friend of mine who met a bear in this same way and the bear came for him and he had to "let him have it."

My bear rose up on his hind legs, and he looked big. I tremblingly stood my ground. He dropped down on all fours again and, after a long, sober look, came on toward me, but only at a walk. I reversed my direction and matched his stride in, I hoped, a casual manner, taking care that I did not give him the impression that I was fleeing, or that he was chasing me, but rather that we were out for a walk together, and perhaps were going to the same blueberry patch! Small consolation in that, if one remembered the jingle:

> There was a young lady of Niger
>   Who smiled as she rode on a Tiger.
> They came back from the ride
>   With the lady inside
> And the smile on the face of the tiger.

I glanced over my shoulder occasionally to see if the situation was changing. Ahead of me was the main ridge, and there we could go our separate ways. At last the narrow passage came to an end, and I turned left on the sheep trail. When the bear arrived at the junction he also turned left. Now it appeared he was following me. A knob put me out of sight and I made a run, dropping back to a walk before the bear came in sight. After a couple of hundred yards I made another left turn down on a spur ridge. I was soon out of sight, and waited. No bear appeared; after a discreet interval I returned to the main ridge — the grizzly was far down the slope! The incident showed that this bear was willing to keep the peace if a fellow would get out of his way.

Like so many happenings, bear experiences come unexpectedly. On one occasion, while we were coming up East Fork River, my companion was following the gravel bar ahead of me. I had been delayed watching a porcupine feeding in a patch of tender fireweed sprouts. To catch up, I cut across a sedge flat in a bend of the river. Suddenly a grizzly loomed up in front of me as though by magic. What had at first appeared to be perfectly flat terrain was broken by a shallow ravine that had hidden the bear until I was too close for comfort. I had only my hands for protection and nothing taller than a grass blade to climb, for this was a treeless country, above timberline. The bear was surprised too, and he rose on his hind legs for a better view. He looked as tall as some of those bears in illustrations I have seen, and I felt as small as the man in the same illustration, but less mighty. It is a most helpless feeling for one to have so little control over a situation and to be completely at the mercy of a bear. I stood there, and so did the bear. I got to thinking that the bear might not know for sure what I was and might decide to come closer to investigate. With this in mind, I took a few slow steps forward, at a diagonal. He watched me closely, and when I stopped he dropped down on all fours and stood facing me. Was he going to come, and if he did, what should I do?

To my relief, he moved into the ravine out of sight, and when he came into view again he was galloping across the river bar, with a great splash every time he crossed one of the many river channels. The tension was completely gone, and I felt a little let down, as though I had been scared when there was nothing to worry about. The behavior of this bear was typical — they usually do run away.

One day in August, when the scarlet buffalo berries were ripe and the first faint touch of fall was in the landscape, my companion and I had a rather unusual meeting with a grizzly. We had been botanizing on Muldrow Glacier, which comes off Mount McKinley, and after gathering a number of specimens we decided to walk over to a small relief cabin, not much longer than a bed, which we had not yet examined. One likes to become familiar with a cabin for several reasons, not the least of which is curiosity.

As we walked toward the cabin we saw a grizzly a few hundred yards behind it, standing erect with his back against a pole and giving himself a back scratch. The scratching taken care of, he started in our general direction, then veered into the willows. When we were about twenty-five yards from the front of the cabin we again saw the bear, now about sixty yards behind the cabin. Twice he stood up on his hind legs, looking sharply around him in the willows for a dodging ground squirrel. We stood engrossed when suddenly the

We ran for the cabin — we, the ground squirrel, and the grizzly!

squirrel emerged from the willows running, with a good start, for the protection of the cabin. The bear was a little slow in spying him but was soon in hot pursuit toward that little cabin. We had but a moment to take stock of the situation. We had no time to get away, yet we were not sure if the cabin door was unlocked. We ran for the cabin — we, the ground squirrel, and the grizzly! And we all arrived at about the same time, the ground squirrel far enough ahead of the bear to escape his jaws.

While the bear was at the rear of the cabin, disgruntled over losing the squirrel, we were at the front door, frantically trying to open it. A storm door, flush with door casing, was held in place by four bolts, one near each corner, sticking into holes in the frame that had been bored in at an angle. It looked formidable, but each bolt came out with a single jerk. There was no door knob, but we found a providential knothole large enough for inserting a finger and pulling the storm door loose. The inside door then confronted us, but happily it was not locked and we were as safe as the ground squirrel. I cautiously stepped outside the door and,

Soon we saw him walking slowly toward Muldrow Glacier, swinging his head our way with each slow step, and growling.

looking around the corner, saw the bear coming along the wall. He stopped short, woofed, and made a stiff-legged jump forward, pushing his front feet down hard on the ground, a bluffing gesture. He was only a few yards away, and I dodged back into the cabin. Soon we saw him walking slowly toward Muldrow Glacier, swinging his head our way with each slow step, and growling and giving us a surly look each time his head swung toward us. We watched him stop along his way to feed on the bitter buffalo berries, and we made a pot of coffee and talked about our little adventure.

6. The Grizzly and Wags's Fish

IN THE SUMMER OF 1940 we were living in a relief cabin on East Fork River with Gail, then four, and our younger child. I was studying wolves and mountain sheep and was out each day, leaving the family to fend for itself during my absence. One day, as I approached home about noon after some early-morning observations at a wolf den, I saw two men, a big grizzly, and my wife outside the cabin. When I drove up, one of the men told me with a twinkle in his eye that as they were passing on the road they saw my wife chasing a grizzly, brandishing a stove poker! The grizzly had first been discovered when he appeared at the cabin window, and had been scared away with much noise and the poker. He had returned, however, and when I drove up he stood fifty yards from the cabin, very much out of sorts, for he was chomping his jaws and foaming at the mouth. This was one of the "spoiled" bears that had become accustomed to people and were unafraid. Up and down the road he had been breaking into food supplies at the road camps, but this was the first time he had come as far east as our cabin. Perhaps we had not been molested because we had burned our garbage and empty cans to destroy all odors that might attract a bear.

He finally recognized our superiority in numbers and we managed to chase him out on the river bar.

That night we were awakened by the whining of Wags, our captive wolf pup, who was chained near the cabin. Her parents and aunts and uncles knew Wags's whereabouts and occasionally came to her during the night, and when we heard the whining we suspected that one of the adults was paying another visit. I opened the door a crack to investigate, and when I turned on the flashlight I discovered, less than ten feet away, a wall of bear hide, the hindquarters of a grizzly protruding from the coal bin. He was after a sack of dried salmon we had recently obtained for the wolf pup. The grizzly was too close for comfort. The door of our cabin was hung flimsily on weak hinges and would be a poor defense if the grizzly decided to come in!

A noise is a good weapon in such situations, and, with a piece of stove wood, I set up a loud racket by pounding on a large dishpan. The bear, greatly startled, left with the sack of fish but dropped it in the path as he ran into the darkness. After peering into the same darkness with the aid of a flashlight, I stealthily gathered up the fish and threw them on a platform cache. We waited for some time to see whether he would return, but there was no further disturbance during the night.

When I drove away in the morning I gallantly took the sack of fish with me to remove this bear attraction from the cabin. An unafraid grizzly around camp was cause for some worry. About four in the afternoon I parked my pick-up fifty yards or so below a road-camp garbage pile which had been burned a day or two before when the camp was moved away. Leaving the car, I followed a little creek, bordered on one side by steep cliffs leading up to the highway and on the other side by a broad, open flat. Suddenly, ahead of me, a grizzly emerged from the willows and came walking rapidly in my direction with long, swinging strides. The camp bear was on his way to the west again. The only place I could retreat without being seen was into the cliffs, but they were so precipitous it was doubtful if I could find a passage upward. However, with some scrambling I managed to get up a short distance before the bear appeared below me.

I remained quiet until he had passed, then maneuvered upward until I reached the highway. From there I again caught sight of the bear, now busy pawing around the cans and debris at the dump. There was little left of interest, and he continued westward, still seeming to be in a hurry and impatient. I was glad he was leaving, for I could get back to my car, parked near the dump. But there was to be some delay, for the grizzly, in leaving, walked directly into a stream of fish scent which had its origin in my car. He stopped abruptly, turned his head to one side, and raised his nose upward, sniffing the delicious, fishy breeze. Clearly, he reasoned, this scented stream of air must come from the dump, for that was where he had always found food before. But how had he missed those fish at the dump?

Instead of following the fish scent back to the car, as he would ordinarily have done, he followed his preconceived idea that all food in this area was at the dump. Back among the cans, he pawed and searched, sniffing in all corners, unable to understand why he couldn't find those fish. He would start to leave, look back at the dump, then return to double-check and paw some more among the cans. Finally, with much hesitation and reluctance, he again departed.

Once more he walked into the stream of fish scent. No hesitation now, but back to the dump at a fast walk for more pawing and looking and pondering about fish. A third, fourth, and fifth time he left and returned. Then, instead of going directly westward on leaving, he followed the side road to the north almost to the highway before he turned west. I felt relieved that he had taken a new route, and I was confident that he was really going this time and would be out of range of the fish scent. I watched him getting farther and farther away; already he was almost twice as far to the west as he had yet been. But no. Even at that distance the breeze carried the fish odor, and the bear recognized it the moment it touched his nose. Now a longer walk back, and more probing among the cans.

After looking for some time, he caught a stray whiff direct from my car, and warily he circled toward it, stretching his neck and lifting his nose. Now at last he had the fish located. This pleased me, for he could get his fish and be off, and I could go home for a late supper. Then I heard a motor coughing. That was a surprise, because traffic on this road, which was not connected with any other in Alaska, was extremely light. The coughing sound approached for some time before I could see its source, an old Model A Ford coming slowly at about ten or fifteen miles an hour. It was Johnny Busia, trapper and prospector from Kantishna, returning from one of his annual or biennial trips to the railroad for supplies. I stopped him, pointed out the bear, told him of my predicament, and asked if he would drive me to the car, toward which the bear was slowly moving. "Yes, sure," said Johnny excitedly, "but you drive." Johnny was not an experienced driver, a speed of fifteen miles an hour on the main road taking all his attention and nervous energy. We made the turn to the dump, and the bear moved off a short way. I had again rescued the dried fish and resourcefully protected the family for one more day.

One of the most exciting and unusual incidents in connection with grizzlies happened in 1949 on a long slope above Ewe Creek in McKinley Park. A geologist and two assistants were climbing the slope. Jack, a well-built, active fellow, was a hundred yards or so ahead of his companions when he saw just ahead of him two patches of light-colored fur. He could not identify the fur patches, but said he thought of mountain sheep. Turning toward his companions, Jack motioned to them to be quiet. Then he looked again toward the fur patches and was confronted by a charging bear. He barely had time to side-step the charge. The bear turned and charged again, and once more Jack side-stepped like a practiced matador. After escaping the second charge, Jack ran downward toward a cliff. Just as he was about to jump over the cliff, the bear was again upon him and raked claws down his packboard and into the small of his back. Jack went over the cliff with the blow, landing in the brush fifteen or twenty feet below. The bear's momentum carried her down the slope to one side of the cliff. This all happened in a matter of seconds, but now the companions came up and succeeded in scaring the bear away by shouting. The wounds bled considerably, but when I saw them a few weeks later, they were perfectly healed.

For pluck and action it is hard to top this encounter — and two witnesses verified the story. I do not know the age of the second of the two patches of fur Jack saw, or anything of its actions. Apparently it was a cub, either a yearling or a two-year-old.

A short time later, another incident took place that began in much the same way. Two men were hiking across rolling tundra when one of them, on the crest of a ridge, saw a patch of fur in the grass below and a leg waving. Thinking it was a young caribou calf, he motioned to his companion. On a closer look, they found it was no calf but a mother bear on her back, playing with two large cubs. The mother, some sixty yards away, saw the would-be photographers. The men moved away out of her sight over the knoll, and the bear, followed by her offspring, came up to the place where the men had been. By this time the two men were running their fastest toward a little pond. One of the cubs, probably following the instincts of the chase, galloped after the fleeing quarry, but the mother and the other cub came only at a fast walk, the mother keeping tabs on her galloping offspring. Circling the end of the pond, the excited men stopped and shot the cub as it galloped toward them. The mother and the other cub departed.

Another exciting incident happened in the Mount Mc-Kinley area a few years ago. I heard the story from the weatherman, who was out on a vacation taking pictures of the mountains and sleeping on the open tundra in his sleeping bag. One morning he awoke to find a grizzly standing over him. Without getting out of his bag he rolled away from the bear a few yards, crawled forth, and retreated with his packsack, which contained his equipment and, as I recall, some of his clothes. The bear by this time had recovered from his astonishment at this unusual experience and followed the weatherman. The man threw the packsack at the bear, who picked it up in his mouth and carried it away, strewing the contents in his wake. The bear disappeared in the willows, but to the weatherman's dismay, it was not long before he returned. The weatherman had no other packsack to serve as a second offering to the bear, but he had a frying pan, all the weapon he needed. Banging on this, he set up such a clatter out on the tundra that the bear was glad to get away. Most of the camp outfit was recovered.

A section man on the Alaska Railroad told me about an incident that happened along the railroad when he was working on the section at Broad Pass. One of the men had left the crew and gone out over the open flat a short distance. Soon the men saw him racing toward them, and they thought he was coming to announce the approach of a train. Then they saw a bear close on his heels. One of the men, a fat fellow weighing 250 pounds and with short legs, cried, "No good to run, boys, we better stay right here!" The man being chased reached the crew with about five yards to spare. A crewman picked up an iron bar and beat on the track with it. This frightened the bear and he "hit off for the tall timber," which means he ran a long way, for there were no trees in sight.

One of my own experiences with grizzlies took place at Wonder Lake, near the base of towering Mount McKinley, where a five-room bungalow was maintained as a summer headquarters for field parties and park personnel. In the spring of 1948, when the roads were free of snow and open for travel after being closed all winter, we found that a grizzly had broken in and played havoc. The flour bin had been carried into the dining room, and the rest of the groceries in the kitchen and pantry were strewn over the rooms and up and down the hallway. Doors were torn off cupboards, the kitchen range pulled out into the middle of the room, the coal bucket smashed. The bear had gone into the basement and feasted on scores of boxes of surplus Army chocolate bars; uneaten ones lay scattered over the cement floor. A bucket of brown paint had been spilled and tracked about the place. The basement windows were all smashed. To reach the basement the bear had broken through a wall between the hall and the stairs. In the dining room he had looked out of two of the windows, for a perfect print of his nose was left on each. He had used a window leading off the front porch into the front bedroom for a door. This was the first time in almost ten years that bears had disturbed or entered this building, but it was not the last.

One day two road men stopped as they were going past and cautiously walked into the house to view the damage. They had looked over things upstairs and began to descend the basement steps. They stopped and listened, but all was quiet. Just to be sure, they threw a stick of wood into the murkiness, across the basement floor. There was a clatter as it hit, then a cough and a commotion. One man, in his flight, gashed his face against part of the building. Out the front door the two men exploded, and close on their heels was the grizzly. They made the car, and the bear fled across the tundra.

The building was repaired, but the bear could not forget those chocolate bars. He ripped part of the siding off the outside wall and made a hole through the wall into a back bedroom. Reaching in with his paw, he pulled a mattress off a cot and out through the opening. And he entered the house through the porch window.

Occasionally I did field work in this area and stopped overnight in this house at Wonder Lake. Once in early September, when the long Arctic days had passed and the nights were dark again, I parked the car in front of the house, carried in a few groceries, and cooked supper. After writing some of my wildlife notes for the day by candlelight, I brought in my sleeping bag and placed it on a cot against the wall opposite the hole in the wall. This cot still had its mattress; it would have been too drafty to sleep on the other, and besides I did not want to be mistaken for a mattress and pulled through the hole. By the front window which the bear had used for his door, I put up a breastwork of furniture and boards in such a way that they would topple over with a clatter if the bear should try to enter. I closed the door from the hallway into the front bedroom, so that if he should enter he would still be locked out of the hallway off which my room opened.

I went to bed and slept peacefully. Near midnight, while I was in the sound sleep of the early part of the night after a day of hiking, a bear was hurrying across the tundra with visions of chocolate bars dancing in his head. Up the front steps to the porch he padded, and then to his window doorway. There he encountered my obstructions, which he pushed aside so gently that they did not fall — there was no clatter! He was now in the front bedroom. I still slept peacefully. Now the bear found the door into the L-shaped hallway closed. He walked about the room, scratched some at random, then concentrated his attack on a panel. In a few minutes big chunks of wallboard were torn loose, and soon a hole was big enough to allow him to pass into the dining room beside the fireplace. He did not come the few steps down the hall to my bedroom but sat down in front of the kitchen door. With his powerful paw he wiggled the doorknob, and soon I started hearing the rattle in my sleep. I awoke and heard the fumbling at the doorknob. I wondered who was trying the door, and remembered that two road men had passed in the evening when I was having supper; they were probably on their way back and were stopping for a snack. I had better get up and see them.

I reached for my trousers and shirt, and the rattling continued. I was about to call to my visitors when I heard a heavy thump on the floor. This startled me and I wondered if it could be the bear. More heavy thumps left no doubt about it. I finished dressing in a hurry. Cautiously I opened a window and peered into the darkness to see if a dark bulk lurked in the shadows, because there was evidence that at least two bears, a family, had cooperated in the first raids.

All seemed clear and I dropped the ten feet to the ground, reached the car, and switched on the car lights to shine on the front porch and the window-door; there I waited for the bear to come through the window. There were orders to shoot this particular bear because of his continued destruction, and I pointed my rifle at the porch and tried the sights. I could aim well enough, and all was in readiness for the bear's appearance.

Suddenly there was a terrific crash from the opposite side of the house, which sounded as though all the siding on one wall was being ripped off at one time. Thinking that another bear on the premises must be active on the outside, I doubled my vigilance. But by one A.M. nothing further had happened and there were no more sounds. I maneuvered the car to the other side of the house, and there I saw the cause of the crash. The bear had jumped through the dining-room window — there were his tracks, upside down on the siding below the window. True to bear tradition he had left through a window!

A week later six people spent a night at the same house, occupying all three bedrooms. They knew about the bear that had disturbed my sleep, but they bravely retired, feeling safety in numbers, though a little nervous. Every time someone turned over in bed a cot would squeak, and six alert sleepers would awaken. No bear came, but many an imaginary grizzly prowled in the hallway that night, and eyelids were heavy in the morning.

That fall, after the snows blocked the passes, the bear

again entered the house. He made a thorough search for the food, which had all been removed. He broke through one wall after another. He entered a bedroom through the door, and also entered it through a wall. The following spring, seventeen panels of eight-by-ten-foot wallboard were needed to repair the damage.

Bear photographers have a safe way of taking bear pictures. They look around for a companion and size him up carefully. There must be no mistake in the choice, for a mistake may be fatal. The choice must fall on a fellow slower afoot than the photographer. There lies safety.

A photographer friend, young, active, and just starting his career, invited me to join him in photographing a mother and large cub he had spotted. I was not too offended at being

The cub stood erect on his hind legs, front paws on the tripod and camera.

invited, because another man with legs far shorter than mine
had already been recruited. Knowing the photographer's
eagerness to get good pictures and therefore the likelihood
that he would approach too near the bears, I turned down
the invitation but watched from a distance.

The two men approached the mother grizzly and her cub.
At first the bears were asleep, but later they grazed, moving
toward the photographers. The professional had an expensive Swedish still-camera and tripod. On the grass to one
side was a high-priced new motion-picture camera. Without
warning, the cub loped toward the photographers. The men
were taken unaware and fled in haste, leaving all the equipment behind except for the camera which the companion
carried in his hand. The cub stood erect on his hind legs,
front paws on the tripod and camera. The tripod swayed one
way and the other, and the animal appeared to be taking pictures. Several times he sniffed around the ground and stood
up again. The mother then came along and also sat up at the
tripod. She examined the tripod and camera thoroughly several times. The camera dropped forward on the tripod, startling her, but only briefly. She now examined the packsack and
movie camera on the ground. This camera fascinated her,
too, but in a different way, for she lay down and rolled on it,
evidently liking the scratching effect of a knob. After an in-
terminable period, during which the two men shouted from
afar, the bears moved away, but the cub returned once more
and again tipped the tripod far to one side. When the bears
had moved off in their grazing, the cameras were retrieved.
The chrome on the still camera was dented, apparently by
teeth, but there was no other damage except to the chewed
binding of a notebook used for recording movie shots.

One day my daughter Gail and I set out to climb to a

band of Dall sheep on the ridges above us, in order to take notes on the lamb crop and take pictures. At the start we followed a narrow, dry stream bed to avoid pushing through willow brush. As we climbed we caught an occasional glimpse of the distant sheep, and after a time we noticed they were all alert. They might have been looking at us, but I felt certain that our appearance was not of sufficient importance to arouse such intent interest. Their fixed look was directed, it seemed, at something a little to our right. I told Gail that perhaps we should get out of the stream bed for a look around, and we scrambled up the bank to an open area and looked to our right across the stream bed.

To our surprise, a grizzly was looking at us over a low clump of willow brush — much too close for comfort. I whispered to my companion that we should back away slowly and she should move toward me, but slowly, with the rifle — I had been loaded down with equipment, and when we divided the load she had taken the rifle. Suddenly the grizzly started moving at a lope around the willow brush toward us. Was he actually charging us? I thought a shot in the air might scare him — but also it might aggravate the issue still more. (Shooting *at* a bear at such close range would be extremely dangerous under any circumstances.)

The grizzly didn't stop at the creek bank as I had hoped he would, but after jumping into the creek he did stop. Perhaps he had gotten our scent and realized that we were humans. Anyway he stood tense, facing up the creek bed. I guessed he was probably half mad and half scared, and I wondered if things were not somewhat in the balance. He was so close to us that he was probably afraid to leave and, in retreating, expose his rear. Then, suddenly, there was action. His hindquarters went down, like those of a dog flattening his rear to avoid a kick or clod of earth, and the bear started up the creek at a gallop. The next time we saw him he was far up the mountain, still galloping, and when he went over the distant ridge top he did not even stop for a last look.

When we first appeared on the scene, the sheep in the cliffs had been gazing intently at the bear. Now, a bear feeding or moving over low slopes in the usual manner does not arouse sheep. Something unusual, therefore, must have concerned them. I learned later what it was. While Gail and I were climbing the slope via the dry stream bed, the rest of the family were botanizing in a patch of spruces, where we had had our lunch together. They had met a bear at close quarters and everyone received a scare. The bear galloped up the slope toward his encounter with us as we climbed out of the gully, and it was this galloping bear that had attracted the attention of the sheep.

One more little incident will show the degree of circumspect watchfulness one acquires in bear country. Three of us started early to hike to the glaciers at the head of one fork of the Teklanika River. The caribou herds cross the Alaska Range at times by way of these glaciers and we wished to observe the trails they had made over the snow and ice during a recent trek. As we crossed the divide from far up Igloo Creek and started down the rolling slopes toward the Teklanika River bars, we were diverted by the flora. Here some hitherto rare flowers were common, and we found a species or two which we had long been seeking. Although our long trip required that we hurry along, we now examined closely the flowers at our feet, as though we were grazing creatures that had come upon a choice meadow. We were

Wags, the wolf pup, and companion. *Photo by the author*.

Wags, the captive wolf, was always friendly and playful. *Photos by the author.*

Dall sheep (*top and bottom*) rams. (*Center*) Ewe and two lambs. *Photos by the author.*

Dall sheep in idyllic surroundings. *Photos by the author.*

(*Top*) Caribou bulls sparring. *Photo by the author.*

(*Bottom*) Bull caribou in early fall. *Photo by Charles J. Ott.*

(*Top*) Bull caribou, August. *Photo by Charles J. Ott.*

(*Bottom*) Gail Murie and a one-day-old caribou calf. *Photo by the author.*

Caribou in July, during their western migration, on gravel bars of Thorofare River. *Photo by Charles J. Ott.*

enthusiastic botanists and were all concentrating. Then Gail remarked, "There's a bear."

Directly in our path, in full sight on the open, grassy slope, was a sleeping grizzly, so close that we hardly dared whisper. We retreated cautiously, but before we had gotten very far the bear looked around and saw us. He looked, identified us, and galloped to a point near the top of a low ridge. He stopped to watch us, then rolled over on his back. From his reclining position he could still see us, but once he grabbed his hind legs in his arms so as to raise his head for a better look. He was just a big, easy-going Teddy bear.

Bears and bear stories are a colorful part of our culture, a resource to be cherished. May the bears continue to roam the hills in freedom, living in native haunts, grazing vegetation like caribou, or, at times, eating a caribou. For the future we must rescue large tracts of wilderness bear country so that we may get away from the humdrum and add zest to living by sharing adventures with bears. And may the true Alaska Sourdough continue to have wilderness in which to roam and in which to measure big bear tracks with his size-twelve shoe pacs.

7. Nokomis and Old Rosy

O~N A GREEN SLOPE~ above timber, surrounded by rugged peaks and ridges, a grizzly mother was feeding on grass — not any kind of grass, but a juicy-stemmed kind (*Arctagrostis*) that grew erect in the moist spots. She also had a taste for a few other plants, such as horsetail, boykinia, and dock, for this was June, when bears feed on this type of food. Earlier in the spring she had been digging roots; later she would turn to the big berry crops.

We called this mother bear Nokomis, a name that goes back a dozen years in the annals of Sable Pass, where this bear had taken up residence. At that earlier time a trapper saw on the pass a mother grizzly with a lone spring cub, and he said it was a scrawny, sickly-looking cub. (All spring cubs are tiny and angular the early part of the summer and have sloping hind quarters which seem underdeveloped, especially when compared to the proportionately wide, solid hips of the mother.) The trapper called this bear Nokomis. Since then we have applied the name to many mother bears on Sable Pass, even though Nokomis, according to Longfellow, was primarily a grandmother.

This 1950 Nokomis was of special interest because she was followed by three cubs rather than the more usual one or two. They were indeed tiny and looked as scrawny as the trapper's "sickly-looking cub." But they raced over the slopes as though they were in the best of health. In this open country, play resources were somewhat limited, and no opportunity for variety in their games was passed up.

One day, when the mother was feeding near some specially large hummocks, the cubs dashed in among them, climbing, sliding, and pushing. Every little rock the cubs met, they climbed many times, and when they encountered a stunted willow they could hardly tear themselves away and were left far behind the grazing mother. They were not always playing, for their growing bodies required that they take frequent naps, and sometimes they just sat among the white anemones, picking at the vegetation. One cub sprawled out on his back as though asleep, and in a minute he must have had a nightmare, for he rolled off the hummock, legs grabbing the air.

One day two of the cubs out exploring were greatly startled when a ptarmigan burst into flight a few feet from them. Twice a short-eared owl left off its mouse hunting to alight on a tussock and watch the cubs for a few minutes.

Often we had the rare opportunity of seeing the cubs nurse. The feeding periods were rather irregular. The shortest interval between feedings was a little over an hour, but it was sometimes three or four hours and longer. The mother might decide the lunch hour by sitting down, or the cubs would remind her by closing in expectantly. Once I saw a cub following its mother and complaining bitterly about its hunger with hoarse barks and whines. She did not always respond immediately to the hints, but usually it was not long before she would roll over on her back so the cubs could nurse. She lay on her back with feet in the air, two cubs nursing from her breasts while the third one attacked one

Twice a short-eared owl left off its mouse hunting to alight on a tussock and watch the cubs.

of the two inguinal teats farther back. She often had one arm wrapped around a cub, and she would raise her head so she could watch her family as they fed. The feeding periods lasted from three to five minutes; they would end when she either rolled over on one side and took a short rest, or almost immediately got up and resumed her grazing. Only twice did I see a mother nursing her young from a sitting position. (The nursing periods and intervals of yearlings, and even two-year cubs, so far as I noted, is similar to that of spring cubs.)

In the year of which I now write, we first saw Nokomis on

June 15. On the twenty-sixth I learned of another grizzly in this area, who was followed by two spring cubs. I thought this new bear had accidentally wandered into Nokomis' territory and would soon be leaving, but she stayed on all summer, and we had two bear families occupying a rather restricted range. Neither family was daunted by the presence of the other. However, there was no fraternizing, and when they found themselves inadvertently near each other, one or the other or both mothers would move away some distance.

This new bear we called Old Rosy, after a supposedly monstrous female who dwelt in the Savage River area many years ago and who, because of her size, was said to have scared hunters and prospectors out of her range. This current Old Rosy appeared to be about the same size as Nokomis; her coat had a slightly grayer cast, which was not evident unless both bears were seen at the same time. One of her cubs, the larger, had almost a white collar around the base of the neck.

On July 10 we found Nokomis and her cubs on the long slope leading down from the steeper part of Sable Mountain. Selecting a suitable spot near a tiny creek, we settled down to watch the bears — and occasionally the scenery, for only ten miles to the south was the backbone of the Alaska Range, with a glacier in every valley and pocket. Farther westward along the range, we could see Mount McKinley, looming far above the surrounding peaks.

Nokomis seemed more restless than usual. Ordinarily she grazed rather steadily, but this morning she was rambling about erratically, stopping here and there for a bite or two of vegetation, the three cubs following. Her restless movements suggested that she was seeking a particular species of plant which required some hunting, or that she was tiring of her grass-and-horsetail diet, especially as it was maturing, and she was hoping to find something better.

Facing up the slope, Nokomis raised her muzzle the better to catch a scent she had detected. The scent had the effect of a mild repellent, for she fed away from its source.

The cause of her anxiety proved to be Old Rosy, who soon came into view with her two cubs and fed slowly westward across two or three shallow, grassy ravines and then moved upward to a green bench, where she lay down. The bench was a strategic lookout, for it gave a good view of Nokomis. Twice, when Nokomis undertook a march of seventy-five to a hundred yards, Old Rosy sat erect for a better view, having sensed a possible purposefulness in the movement and therefore wanting to keep up with events.

We had been having some foreboding about these bears, wondering what would happen if they came together. Nothing unseemly had yet taken place during at least two weeks of their joint occupancy of this area, and it began to appear that the natural inclinations of each mother to keep herself and cubs apart from all other bears would adjust matters for them. But still it was evident that the presence of two powerful grizzly mothers living, so to speak, in the same household, created a situation fraught with hazard. The motherhood emotions dominated and might with little provocation release ungovernable fury, bringing into play two powerful arms and crushing jaws. During the day, as we watched, we referred to these possibilities, but apparently everything was peaceable. Nokomis nursed her offspring at eleven-fifteen A.M., at one P.M., and again at three P.M. At four P.M. Old Rosy, high up the slope on the green, also rolled over on her back to nurse her cubs, and then relaxed and lay dreaming about her idyllic surroundings.

The hours passed rapidly. A ground squirrel, having become accustomed to us, came close and stuffed its cheek pouches with a lichen that is a favorite food of the caribou, and went on to taste the leaves of a tall sage. Eagles soared high above the mountains, and in the distance a falcon or jaeger was attacking one of them as it sailed ever higher in great circles, without a wing beat.

The afternoon was waning. It was after five o'clock. Old Rosy was now grazing on the bench where she had slept. Nokomis, lower on the slope, had taken a notion to seek grazing farther to the westward. Her stride was slow and deliberate, huge arms reaching forward and hind legs bringing up the rear. Although her pace was slow, the cubs, because of their play, lagged behind. Nokomis passed a point directly below Old Rosy's position, but she plodded forward, taking no notice of the family above her. Nor did Old Rosy at this time show any nervousness; perhaps she had noted that Nokomis was taking a line of march not in conflict with her own position. The high points of our day appeared to be over, for now Nokomis passed into a hollow out of our view.

A few minutes after Nokomis had disappeared, we saw one of her cubs galloping rapidly over the back trail, obviously lost. We wondered if the mother would realize her loss and come back for the cub, but several glimpses of Nokomis in the hollow showed that she was still moving westward, seemingly unaware of anything amiss. When directly below Old Rosy, the cub stood up to look around. He may have cried out, but we could not hear him from our position. Now Old Rosy had stopped her grazing and was looking intently at the tiny black cub below her. What she saw probably registered in her mind as "lost cub" . . . "my lost cub" . . . "cub in distress."

She should have counted her own two cubs, but she didn't. They were behind her, and she was concentrating so hard on that one below her that she failed to check over her own family. Furthermore, her arithmetic may not have been very good, and maybe it would not have helped matters if she had looked around. Of course, I do not really know what Old Rosy was thinking, but things were registering, and rapidly.

Suddenly it all broke into action. At full gallop she started down the slope toward the cub, each long jump charged with emotion. When the cub saw Old Rosy galloping, he fled in terror and, as it happened, in the right direction, toward his mother.

At the start, Nokomis was not aware of her lost cub's plight, but now she saw the desperate situation and, rescue bent, galloped back toward him. When they met she turned and ran on with him to the other two cubs, and all four fled together.

The chase went out of sight in a depression where Nokomis was probably overtaken and forced to face Old Rosy to protect her cubs, for in a few minutes the three cubs emerged alone from the hollow and galloped halfway up the talus slope of a cone near the base of Sable Mountain.

Soon the mothers were in view, and we saw Nokomis maneuvering to keep herself between her cubs and Old Rosy. The three cubs had moved upward and were waiting in a row on the peak of the cone, watching and ready to run again. Old Rosy, who could not be prevailed upon to give up the chase, continued to press forward even though met by Nokomis at every turn. On the steep slope of the cone the mothers stood near each other, facing the same way, shoulder to shoulder, a position two bears take when on guard. They were sparring, holding back, not quite willing to fight. Thus they moved slowly up the cone.

The cubs, trying to keep away from the danger, dropped down on the other side of the cone and climbed some distance up the main slope of Sable Mountain, where they halted again to watch.

After moving farther up the steep slope, the mothers suddenly faced each other, jaws wide open. They rose up on hind legs briefly then dropped down on all fours and closed in, open jaws meeting. Old Rosy had managed to get above Nokomis and now forced the attack by making short rushes. Nokomis met the attacks with jaws wide open, but was forced to keep backing down the talus slope with each fresh onslaught. After this encounter the mothers walked slowly upward and around the cone again, as before, shoulder to shoulder. But in that last skirmish Old Rosy appeared to have gained a psychological advantage. She now was on the uphill side, and after they had gone on around to the back side of the cone, Old Rosy went forward alone and Nokomis moved about near the cone as though she were unaware of the plight of her cubs above her. I could not understand why she made no further effort to halt Old Rosy, who now had apparently sighted the cubs and was galloping up the slope toward them.

The three cubs, traveling close together, saw her coming and did their best, but we saw them overtaken near the top of the mountain after they had climbed over 1,500 feet. When Old Rosy closed in on them and pounced on one, the other two escaped, one galloping wildly down the mountain and the other upward and over the summit.

We could not see what Old Rosy was doing with the cub she had captured, but when she left it, there was no movement. Now Old Rosy galloped over the top after the cub that had gone that way. Shortly she came back in view and moved on a diagonal across the slope, directly toward her own two cubs, which were still on the bench where she had left them two hours and thirty minutes earlier.

In the meantime Nokomis was wandering about on the lower slopes of the mountain, obviously searching for her family. She had missed seeing the cub that escaped down the slope and did not find it for some time. When the cub finally saw its mother approaching, it must have recognized her, for, though somewhat diffident, it did not run away. They touched noses, and the mother nosed the cub all over.

Nokomis was aware that she was still missing part of her family, for she continued searching. Her mouth was open, and foam covered her lips. The cub, after its violent exercise, had no energy for play and followed the mother closely. They backtracked to where they had been in the early part of the day, searched the area, then wandered back toward the talus

The cubs, trying to keep away from the danger, dropped down on the other side of the cone.

Old Rosy had managed to get above Nokomis and now forced the attack by making short rushes.

cone. Old Rosy, feeding with her two cubs, saw Nokomis passing below her and hurried away to higher ground. When we left, about nine P.M., Nokomis was hunting in the vicinity of the cone, and she had not eaten since five o'clock. In reviewing the incident, we assumed that the first captured cub had been killed, but we did not know the fate of the one that went over the top of the mountain.

To check further on the lost cubs, we returned to the scene the following morning. Nokomis was feeding restlessly and still had only the one cub. She made a long trek over to the cone, still searching. We noticed three or four eagles circling over Sable Mountain and alighting on the ridges. Their presence was ominous. They were interested in the area directly below the point where the first cub had been captured. Apparently the body had been dragged down the steep slope among the boulders. In a draw near the summit of the mountain an eagle lit on a dark object and commenced to feed. After he had flown away, a second one sailed into the draw and fed. This one tried to fly off with the remains but only succeeded in dragging them ten or fifteen feet farther down. Then a third eagle fed, and when it was finished it managed laboriously to fly off with what remained of the cub, but crossing a low ridge, dropped its load. These remains were those of the cub that had climbed over the top of the mountain. Old Rosy had probably killed it before returning to her own family. Later we climbed the mountain and found bones and hair of the victims.

In explanation of the tragedy, it appears that Old Rosy thought she was rescuing a lost cub and then, unexpectedly, overtook cubs with their stranger scents. Her violent reaction to these is probably normal grizzly behavior. Two bear families on Sable Pass seemed to have created an environment hostile to the cubs. Perhaps among grizzlies two families in an area is the beginning of congestion, which tends to create its own balancing forces.

The two mothers continued to live in the area for the remainder of the summer and managed to avoid further serious complications, but once in late summer I did see them at rather close quarters.

After filling up on blueberries and crowberries, Nokomis and her cub moved into a green hollow, where they fed for a short time on a tall saxifrage (*Boykinia*) and horsetail. Then she took a notion for a lark and made a chase for ground squirrels and perhaps a search for better berry patches. With long strides she started walking down the hollow, then broke into an easy gallop. Her cocked ears gave her an alert expression, and she kept a close watch ahead as she galloped, apparently on the lookout for unwary ground squirrels.

Her progress was heralded by the sharp warning calls of the ground squirrels, who stood erect by their holes. Those in the direct line of travel dived underground; those off to one side kept their stations and repeated the warning calls. In spite of the signal system, there were always some squirrels to be surprised in precarious situations. Nokomis turned sharply to one side and in a burst of speed almost overtook a squirrel dashing for its hole. She usually caught most of her squirrels by excavating, so she started digging operations, pawing rapidly, both paws ripping the dirt alternately and sending clods between her hind legs, ten feet or more behind her. This time the burrow construction favored the ground squirrel, and after considerable expenditure of time and effort, Nokomis abandoned the enterprise. She went forward as before; after a long, rapid run quartering up the slope, she chased another ground squirrel into a hole. More digging, once more to no avail. Thus they traveled down the draw and over a rise into another draw and out of sight.

On the sky line ridge up ahead, whom should we then discover but Old Rosy, almost buried in a huge hole, only her big hindquarters protruding. Her two cubs idled nearby. And when Nokomis reappeared she was traveling directly toward Old Rosy! Neither bear was aware of the other until they were about a hundred fifty yards apart. Then both stopped to eye each other. In a moment Nokomis retreated, and Old Rosy and her cubs galloped down the slope in pursuit. After a run of about two hundred yards, Old Rosy stopped, stood erect on her hind legs for a long look, and returned to her ground-squirrel excavation. After digging deeper she had to give up, and she then moved over the ridge. Nokomis returned to the draw and grazed on blueberries with her cub. Such meetings were probably not uncommon.

A noticeable change took place in the relationship between Nokomis and her offspring. When she had the three cubs she hardly ever played with them. The three were self-sufficient in that respect and at times did not deign to play with the mother. On one occasion, for instance, the three cubs were playing at a stunted willow, rubbing against and clawing the trunk. Soon the mother came up to them, and now she rubbed against the willow trunk. The cubs all stood aside and watched while the mother rolled over on her back with all four feet waving and kicking in the air. She reached up at

The cubs stood aside and watched while the mother played. Their play had been interrupted.

the willow with her forepaws, clawing at it and pulling down the lower branches. She was in a playful mood, but the cubs stood apart and she played alone. Then she moved on to her grazing and the cubs resumed their play, which they apparently felt had been interrupted.

The three cubs had played by themselves. Now the lone cub sought its mother for play.

But now life in company with the lone youngster was much more intimate. It sought its mother for play and she responded generously. Sometimes the play continued for fifteen or twenty minutes, the mother lying on her back and the cub biting at her feet, head, and neck; sometimes they faced each other and sparred with open jaws. The mother seemed to enjoy the play as much as the little one. Once, after they had played with a broken-down telephone tripod, rolling on the poles and biting at them, they went off a quarter of a mile feeding. In about twenty minutes they walked back to the poles and resumed their play. However, in this case the mother probably had another incentive besides the play, for she enjoyed rolling on the logs to scratch her back. Sometimes she rubbed against the upright poles.

Both families were last seen a few days after the middle of September. As I write, in December, Nokomis and her cub are sleeping in one den, and Old Rosy and her two cubs are sleeping in another. Next summer they probably will again be in their old haunts. On Sable Pass and elsewhere in Mount McKinley National Park, provision has been made for the grizzlies to live their free and natural lives. Nokomis will in due time have a sweetheart again; she will be affectionate and steadfast and, following old grizzly traditions, will have a honeymoon in June. The following year there will be one, two, or three tiny cubs, and then parental affection will be strongly developed and romance forgotten.

8. Why Teen-age Grizzlies Leave Home

WHEN TWO OF US arrived in the green, treeless landscape of Sable Pass on the morning of July 14, 1953, a mother grizzly, followed by a tiny, dark spring cub, was grazing steadily in the rain. About fifty yards away was a small bear; this was a surprise, because a mother bear aims to keep well isolated from all other bears. The arrangement was unusual and we conjectured about it. We looked at this bear with naked eyes and through field glasses and telescope in our efforts to get the best perspective and judge his size and age. Size is deceiving, but the female was near enough to give us some comparison. He obviously was too large for a yearling. After weighing the matter carefully, we judged that he was at least a two-year cub; he might well have been a three-year cub. If he was a three-year cub he might well have been Nokomis' cub, the one that had escaped Old Rosy three years before in this same area. He was a picture of despair as he stood disconsolately in the rain in a forlorn posture, watching the mother and spring cub. The dark hair bordering the eyes and the wet, bedraggled coat added to his mournful appearance.

After a time the sad bear moved hesitatingly forward, advancing ten or fifteen yards toward the mother and cub. Without visible warning, the mother suddenly turned and charged. The bear scarcely tried to escape and was quickly overtaken, and we saw the mother bite him high on a hind leg. She stood over the big cub who lay cowering beneath her. For a minute or two she held a grim pose, looking straight ahead, then strode back to her latest offspring. The attack had been brief but vigorous.

The large cub slowly rose to his feet, sunk in despair, his head drooping.

The large cub slowly rose to his feet, sunk in despair, his head drooping. The hair above the knee became blood-soaked. Obviously he was an unhappy outcast, a cub no longer wanted by the mother, his place taken by another. The cub was undergoing a teen-age crisis; against his wishes, home ties were being broken. A mother that had been ready to fight fiercely for him, no longer cared. In earlier cubhood he had no doubt been cuffed, but this was different, and the truth had not yet dawned on him. As the mother grazed into a hollow, the outcast, with a slight limp, followed at a distance and stood, or sat on his haunches, to watch and hanker for mother companionship.

Meanwhile the spring cub amused itself in various ways,

playing with low willow brush and romping about. Sometimes it approached to within six or seven yards of the bigger cub, indicating a long familiarity. When the spring cub cried, it was nursed by the mother as she lay on her back. A half hour later the cub was nursed again — a much shorter interval than usual — and a third nursing took place in the afternoon after a three-hour interval.

When the spring cub cried, it was nursed by the mother as she lay on her back.

The outcast stood watching from the upper edge of the hollow for over an hour. Then it fed, at first listlessly, then more hungrily. At one o'clock the bears all lay down and rested for an hour, the outcast a short distance apart. When grazing commenced again, I heard the mother growl as she fed, and a few minutes later she again charged the outcast, barely missing his rear with a powerful side swipe of her paw as he galloped away. Later the mother chased him again and he had to flee to escape.

All afternoon the dejected cub remained as near the mother as he dared. Once, when she was resting, he stole slowly to within a dozen feet of her rear and lay down, but then, as though feeling uneasy, he moved farther away.

When I left, about five in the afternoon, the female was feeding her way down a draw, and the large cub was following at a discreet distance. Later in the summer the mother and spring cub put in an appearance on a few occasions, but we did not see the outcast again; he either became sufficiently discouraged to go his own way, or he had been seriously injured before he learned that grizzly mother love cannot be shared by the older generation.

This incident, if the large cub were two years old rather than three, would have been unusual, the result of earlier unconventional grizzly behavior. The possession of a spring cub by the mother would then show that she had bred when followed by a nursing yearling, and this, according to my other observations, is not done in grizzly society. And what breeding takes place by females followed by two-year cubs is not known. I have observed several mothers followed by two-year cubs during the summer that apparently did not mate. More recently I made one observation that indicated that a male may have mated with a female still nursing two-year-old cubs. The male was near the family on friendly terms for a few days, suggesting that a mating period was just coming to a close.

Regardless of the big cub's exact age, it was apparent that an older offspring was not wanted. It seems likely that mother relationships with the teen-ager were disintegrating most rapidly the day I watched, and that a certain mother tolerance was giving way to dangerous antagonism.

Once I observed some bear actions that indicated a family breach brought about by romance. Two bears, one large and angular and the other smaller, undoubtedly a pair, were digging roots together on a river bar on June 9, 1955. Hovering on the outskirts was another bear which, according to my identification, was a two- or three-year cub, probably the former. He moved cautiously toward the female, and when he was about a hundred feet from her, she charged, causing him to break into a gallop to escape. Twice the big male walked slowly toward the cub, but with no obvious intent to harm it, and the cub seemed unalarmed and walked away at a leisurely pace. But whenever the cub approached the female, she charged, and once she chased him for at least a hundred yards. Here again an antagonistic mother was apparently ridding herself of an offspring that was reluctant to leave.

As I have said in another chapter, it is my impression that with two or three offspring in a family there is less dependence on the companionship of the mother than in the case of a one-cub family. With families of two or three cubs, the breach with the mother at any time may be, therefore, more gradual and mutual, with perhaps no one becoming unhappy. I have seen a pair of two-year-old bears moving far off from the mother in a spirit of independence, even though they still followed her quite closely at times.

After leaving the mother, cubs usually maintain a mutual companionship that also eventually breaks down, until one finds them traveling independently and they become lone bears, the normal status of grizzlies except when breeding or raising a family.

The behavior of the two mothers chasing away their teen-agers is similar to what we find in some other species. The mother moose, anticipating the approach of a new calf, tries to rid herself of her one or two yearlings by threats. If the yearling is around after the calf is born, the mother continues her antagonistic behavior. The mountain sheep likewise discourage their yearling offspring from following at lambing time and thereafter, though some yearlings continue to fol-

low the mother and new lamb into the winter, even though
unwanted. Thus we find the pattern of behavior of the grizzly mother, at least under certain circumstances, similar to
other species very distantly related.

9. Picturesque Moose

I N THE DISTANCE the moose looms as a huge, black animal wearing white stockings on rear legs. One's first impression may be that it is rather ungainly. And a close-up portrait of a cow-moose face would win no beauty contest, especially if she has just raised her head after submerging for water plants, with ears hanging and water dribbling from all points. The nose is long, bulbous, loose, and overhangs; the eyes are small, the ears long; the shoulder hump is exaggerated, and the legs are inordinately long. A bell or dewlap hangs from the base of the muzzle that varies much in size and form; broad and shapely in the bulls, but usually fingerlike and only a few inches long in the cows. Each of these accentuated features no doubt serves a purpose. For instance, the long legs are just the thing for traveling in the deep snows of winter, for wading in the water and mud of marshes, ponds, and lakes, and for stepping over high windfalls; in enabling him to reach high in his browsing, they also greatly increase his winter food resources. When all the details of his appearance are assembled, the moose becomes a highly picturesque wilderness animal; and a large bull in the fall, with its spread of palmate antlers is truly majestic.

Apparently the Alaska moose is considerably larger than the eastern and more southern animals. It measures six or seven feet at the shoulders, and a large bull probably weighs upward of 1,500 pounds. The antlers may reach a spread of about eighty inches.

The moose becomes a highly picturesque wilderness animal.

Because the moose is such a large animal, hunters find it a considerable task to pack in a carcass after a successful hunt. Dr. Otto Geist told me how one of his Indian friends at Bettles on the upper Koyukuk River had solved the problem. Big Charlie, the Indian, went hunting and shot a moose. Instead of transporting the moose to his family at Bettles, Big Charlie thought up an easier way. He hauled his family to the moose and pitched a tent over it. There they remained a number of days, feasting. When they returned, Big Charlie said that his wife had a sore stomach.

The moose is especially sensitive to scent and sounds. Scent messages are minutely regarded, recognized, and acted upon. The big ears catch the faintest sounds, and the animal becomes attentive to learn what is moving. Eyesight is difficult to evaluate, but discernment and recognition through eyesight seems quite deficient when compared with those of many species, such as mountain sheep or deer. Also it seems that the eye is not kept as alert as the other senses, which function more automatically. However, the moose notes a moving object, and in approaching one it is well not to underestimate its vision.

Moose are rather silent except during the mating season. At this time the cow occasionally utters a drawn-out, wailing "wuow-wuow-wuow" and the bull's call is a deep, gruff grunt: "ugh." I have heard a bull also utter a whining call when with another bull during the rut. When a moose is alarmed or curious, he sometimes utters a loud bark similar to that of an elk under like circumstances, but even more guttural. One spring in Wyoming a bull that had caught but a brief glimpse of me before I sank out of view uttered at intervals a series of explosive roars that were really startling. He was apparently curious, for he moved about in the willows, not chancing to come closer but trying to see me. Only at one other time have I heard this roaring call. One day, also in Wyoming, I was examining the remains of a red squirrel that a marten had eaten on the snow when I was startled by this blasting roar. The moose was hidden by spruces and apparently had my scent only. At long intervals, for a period of ten or fifteen minutes, this moose uttered his roar, which amounts to a greatly intensified bark.

Frequently one may hear even big bulls making soft, baby-like mewing sounds: "mmm, mmm." I have several times

heard a large bull give this soft call when approached by another. It was difficult to know just what the call signified — it
could have been a warning of displeasure at his companion's
too-close approach, or a mark of recognition. On March 14,
1949, two cows, a yearling, and an old bull were resting in
the woods along Hines Creek. The old bull, lying about a
hundred yards from his companions, rose and cried softly as
though lonesome, which possibly he was, for he kept calling
as he walked to them and lay down nearby. After a time the
cows and yearling moved away to feed, and a little later,
when the bull rose and followed after them, he again made
these soft sounds.

The bulls have a lazy time in summer, alternately browsing and retiring to rest in the cool shade of the spruces. In
1956 three large bulls spent the summer together along Igloo
Creek, confining most of their movements to a stretch of less
than two miles. They were constant companions, and they
often rested only a few yards apart. I saw a fourth bull, an
extremely large animal, a few times in the area, but he
seemed inclined to live alone.

The mating period begins early in September and lasts
about a month. It is presaged by rubbing the velvet off the
antlers. The three bulls at Igloo Creek removed the velvet
on the last day or two of August. They thrashed the brush
and rubbed the sapling spruces until small branches were
stripped off and the slender trunks partially or entirely gir-
dled. One bull tossed a small log into the air and uprooted a
little sapling. Another had his antlers adorned with a spruce
branch that hung down along his cheek. Here and there,
shallow depressions were made by pawing and horn scraping.
On September 3 one of the big bulls at Igloo turned up
with a lame hind leg and showed prong scrapes on his side.

Apparently he had been in an early-season altercation. We found him off by himself on the injury list before the mating season had fairly started. The aggressive spirit of the bulls was developing and there was much sparring. Two young bulls with miniature antlers were trying to spar but had trouble finding each other's antlers. The sparring continues throughout the season, and there are many post-season jousts, as evidenced by tracked-up snow and bits of hair scattered about.

Two or more bulls may move about together in search of a cow, sparring occasionally, pushing against each other mildly, and separating when one or the other gets too vigorous. I have seen a big bull tip its antlers slowly from side to side as it advanced to spar with a companion, and then stop — they had no cow to fight over.

The earliest date on which I have seen a cow being definitely followed by a bull in the park is September 3. On this occasion a cow and a large bull with cleaned antlers were feeding on willow in the tundra a few miles from Wonder Lake. The cow departed down the rolling slope toward Mc-Kinley River, but the bull moved a few steps to the top of a little ridge and watched me for several minutes. Then he followed the cow, which was half a mile ahead of him, and as he traveled he grunted, "ugh," at intervals, as bulls do when following a cow in the rut. Two or three times the bull broke into a trot to overtake his cow.

My observations indicate that a bull usually consorts with only a single cow at any one time. The bull neither herds nor chases the cow; she travels and feeds where she pleases and he faithfully follows. I have seen a bull with two or more cows forced to choose one of them when the cows turned separate ways. One year, on September 19, I saw a bull standing in the midst of four cows that were from a hundred and fifty to three hundred yards away from him. The cows were moving from each other, and it appeared that the bull was uncertain which one to follow. I have, however, seen a bull with two cows and all behaving as though relationships were fully stable.

One year, on October 6, when most of the cows had probably·finished mating, I saw near Savage River at the base of the slope, a loose assemblage of moose consisting of seven cows and six bulls. The cows were scattered but the bulls were close together, very tolerant of one another. Two of the bulls were sparring and a third one was trying to take part. The bulls were still interested in the cows which, however, seemed unresponsive; probably they had already finished breeding. One bull nudged a resting cow, causing her to rise, but after following her a short distance he left her. Later the bulls followed a group of three cows but soon stopped and fed. The bulls were finding the cows no longer interested.

Several years ago, on Isle Royale in Michigan, I had many opportunities to watch the moose during the rut. For a number of evenings I stationed myself in an aspen or spruce tree at the head of one of the coves to observe activities. A cow would usually make her appearance at a salt lick, followed by a bull big enough to fight off all rivals. During the evening other grunting bulls could be heard in various directions. Young bulls were tolerated to some extent, but a large bull was a challenge. The bull in charge would advance toward a challenger with slow and measured step, grunting at intervals and stopping a few times to thrash the alder brush with his antlers or to demolish a sedge hummock. The approaching bull could also be heard fighting brush and grunting. But the challenger always retreated from the bull in charge. At this time I found it easy to bring a bull to me by breaking branches or by uttering the wailing call of the cow. I often enticed bulls in this way almost to the foot of the tree from which I watched. On a few occasions a bull approached me in the woods, apparently hoping that the sounds of my walking came from an unattached female.

The calves, reddish brown and without spots, are born in late May and early June. By fall they have made a surprising growth and have acquired a dark pelage similar to that of the adults. The calf follows the mother all winter, but with the approach of the calving period the mother no longer wants or tolerates her year-old offspring. One or two weeks before the new calf is born she makes short threatening runs at the yearling to chase it away. She generally succeeds in ridding herself of the yearling, perhaps, but occasionally I have seen a yearling following the cow and new calf at a safe distance.

The calves are often left behind to rest, but I have never seen any indication of a moose, deer, or elk "hiding" a calf. Usually the youngster lies down when it feels so inclined. The mother knows where it is and she wanders off to feed, sometimes for a few hours before returning to it. At Wonder Lake, even late in the summer, I saw a mother leave her calf behind often while she went off to feed. Sometimes she fed at least a mile from where the calf rested. On August 3, 1948, I watched the cow and calf feeding on willow leaves about two hundred and fifty yards from the lake. After a time the mother moved toward the lake to feed on water plants. The calf watched the mother as she left, stopped its feeding, and trotted about two hundred and fifty yards diagonally up a slope and lay down in an extensive, heavy grove of alder. About two hours later the mother left the water, fed some more on willows and also on fireweed, and moved slowly along toward the calf. She uttered an occasional prolonged bleat which I could hear from a point two hundred yards away. She apparently did not know the exact location of the calf, for she moved some distance beyond its position and then circled. Then she lay down about a hundred yards from it, apparently having heard or scented it. This alder grove was probably a regular rendezvous for these two.

A friend taking movies in the park said he had watched two bulls feeding for several hours, waiting for them to move into a beautiful mountain background. I asked him what the moose were eating, and he replied, "All I learned was that they eat a lot!" And it is true that this huge animal needs to feed rather steadily for long periods.

The moose is primarily a browser. His long legs make it difficult for him to reach short vegetation, such as grass on firm ground, but this handicap is offset by the advantage of being able to reach high for twigs on tall brush or tree growth. He can trim a tree to eight or nine feet, and in winter he may raise that browsing mark to twelve or fourteen feet, depending on the depth of the snow. When dealing with saplings or tall willow brush, the moose, if necessary, resorts to breaking off sapling and willow limbs to bring the tops within reach. It is said that he rides down the brush between the front legs to get at it, but in all my observations he has either grasped a branch in the mouth to pull it down or pushed it down by placing the muzzle over the sapling and pressing it down until it breaks. Sometimes he pushes the muzzle farther toward the top during the bending, to get better leverage.

In some regions, in the spring, when the sap is running, aspen bark is removed from the trunks in strips several feet

long. In winter it is less easily eaten, but small patches are
often gnawed off, the tooth marks showing plainly. Aspens
felled by beaver in moose country are utilized freely by
moose, both the bark and twigs being eaten.

In McKinley Park the winter food consists chiefly of the
twigs of willow and dwarf birch, of which there is an abundant supply. Alder has been found eaten, in fairly large
amounts in a few places, but in general its palatability in
winter is low. There are over twenty species of willow in the
park, and observations indicate that some species are particularly well liked. A reddish-stemmed species (*Salix pulchra*)
that retains most of its dry leaves through the winter is one
special favorite, and a lush, hairy-twigged species (*Salix richardsonii*) is another. The willows show considerable browse
sign in some areas, but on the whole there has as yet been no
overuse. Many branches of the tall willow (*Salix alascensis*)
along Igloo Creek had been broken down the winter of 1954-
1955, and I noted that rabbits had fed on some of these
broken branches. Green food is always at a premium in winter. I noted in two successive winters that moose had pawed
through snow to get at a patch of grass (*Calamagrostis*), a
species ordinarily of low palatability, to feed on the green
base of the clumps.

In summer, moose continue to feed on willows and dwarf
birch, but chiefly on the leaves and new twig growths. These
two species make up the bulk of the summer food of most of
the moose in the park. Tall herbs such as fireweed and chiming bells (*Mertensia*) are easily grazed. Cow parsnip is a favorite when available. Young grass and sedge in wet places
are often eaten in early summer. Where aquatic vegetation is
present it may make up a large part of the summer diet. Wonder Lake and other ponds in the park are quite often visited

by moose. In seeking aquatic vegetation where the water is rather deep, moose frequently submerge completely. Usually the head is held under water twenty to forty seconds, but on Isle Royale, Michigan, one moose remained submerged several times for one and one-half minutes. A photographer there had a longer record, but when he developed his movie film he learned that the moose had emerged for air once during the timing.

In Teton National Forest, Wyoming, in the fall of the year, I watched a cow and calf feeding on mushrooms growing on hard ground. The calf, because of its long legs, had difficulty reaching the ground and was on its knees most of the time; and the cow, for comfort, was often down on one knee. Sometimes, to reach the ground better, her front legs were widely spraddled. Mushrooms are a delicacy to many animals and apparently the moose like them well enough to go to this extra effort to feed on them.

On the Kenai Peninsula, studies by Palmer and Chatelain showed that a number of moose calves had been eaten by black bears. The remains were found in the droppings. But to what extent the bears had killed the calves was not determined. Such data are difficult to gather. But it is likely that a certain proportion of the calves eaten were healthy calves which the bears had managed to discover and catch. Moose regularly leave their calves alone while they move about feeding, and there is always a chance that a bear may find an unprotected calf. Some of the calves eaten are no doubt carrion, and possibly a few weak calves are taken.

In the park, moose are found in both black-bear and grizzly habitat. Although I have no data on the black bear, a few incidents indicate, as we would expect, that the grizzly is much interested in moose calves. But apparently, if the mother catches the grizzly prowling about near her calf, she goes on the warpath and at least holds her ground. John Williams, a former employee of the park, watched a mother, followed by her very young calf, determinedly chasing a grizzly and doing her best to overtake it. Charlie Ott saw movies taken by a bus driver in the park that showed a bull moose chasing a grizzly for some distance and striking at it. On the other hand, I was told by an eyewitness that a grizzly chased a calf into a clump of brush and that the mother left with only one of her two calves. The grizzly and the other calf were said to have remained in the brush, and the conclusion was that the calf had been captured. No doubt the grizzly captures an occasional calf, but its predation probably causes only a nominal effect on moose numbers in the park.

In Wyoming there are reports of grizzlies killing adult moose in the spring of the year in special situations. It is when a bear finds a moose floundering in deep snow, sufficiently packed so the legs break through at various angles, that the bear is said to prey on adult moose. I have observed moose floundering helplessly in spring drifts. Under such conditions it would seem that a grizzly could easily overcome a moose. Probably snow conditions in McKinley Park would seldom give a grizzly such an advantage.

On May 25, 1955, I saw a cow moose standing in a draw. Her inactivity was unusual, and I stopped to watch. Soon she focused her big ears forward, and I assumed she was looking either at one or two newborn calves or perhaps at a fox up a side draw. Then, to my surprise, I discovered a large grizzly on a steep slope thirty or forty yards directly above the moose. The bear was lying on its side, its huge head resting partly on one paw. It had that Teddy-bear look that bears have when the head is well furred and the muzzle appears small. The bear, I thought, was biding his time, hoping for some kind of favorable opportunity to capture a calf. Or perhaps he had already managed to capture a calf and was sleeping after a big meal, and the mother moose was remaining close to a tragic scene. The bear was safe from a sudden rush by the moose because of the steepness of the ridge and was able to lie close by. I waited and saw the moose cock her ears toward a spot just in front of her; I also saw the bear lift its head and look briefly down at the moose, and then go back to sleep again. Both animals had an abundance of patience. The bear later moved up the ridge four or five yards and stretched out on its side again.

About half an hour after I had started my observations, the bear galloped across two or three shallow draws as it circled to one side of the moose, coughing two or three times as it galloped. After traveling about fifty yards the bear, still about the same distance above the moose, stopped to look at her. Then, after a minute or two of watching, the bear continued on the contour past the moose, dropped into the draw, and continued loping until out of view. Later its progress was indicated by flushed ptarmigan and the running of a startled caribou. I do not know what caused the bear to leave. It may have decided to seek its more usual early-spring root diet. I failed to see the bear rise and so I did not observe its nose action before it started moving, but from past experience I rather doubt that human scent would have precipitated the action.

A few minutes after the bear disappeared, I saw two tiny newborn calves standing in front of the moose! There actually *were* calves and the mother had kept a grizzly from them. After the bear left, the mother moved up the draw feeding, and the two calves followed.

Because the moose is large and formidable, the wolf no doubt prefers hunting smaller game, such as caribou, mountain sheep, and snowshoe hare. But when these foods are scarce, the wolf probably turns oftener to moose for food. If a moose is at full strength and the snow is not so deep as to tire him in his stand, his defense efforts are usually determined enough to discourage the pack. Apparently the power of the wolf pack and that of the moose are closely balanced. An encounter involves danger to the wolves as well as to the moose, for the wolves are sometimes incapacitated by the sharp hoofs of the moose. When the wolves test a moose, they no doubt become aware of any sign of weakness which may become evident and, if faltering is noted, become more persistent in their harassment. The weakness in the moose may be due to old age, malnutrition, parasites, or disease, or it may be due to a condition of deep or crusted snow. No doubt the lone moose is in greatest danger. And it is probably true, among moose as among other ungulates, that some of the lone animals are alone because they are not disposed to move about with healthier animals. I suspect that all predation takes place in winter except when unprotected young calves happen to be discovered in summer.

During the period when mountain sheep were abundant and vulnerable in the park, little evidence of predation on moose was noted. In later years, when sheep became less available, my impression was that predation pressure on moose increased somewhat.

The following incident shows what sometimes happens when wolves encounter a moose. On March 5, 1950, I saw ravens circling over the spruce tops north of Hines Creek near the edge of the woods. Traveling on skis, I followed a draw which had recently been much used by foxes. I suspected that they had been visiting whatever had attracted the ravens. As I progressed, I noted tracks of fox and wolverine crossing my route, indicating that I was approaching a point of convergence. Soon I saw wolf tracks, too, and then I saw that a wolf had been dragging a heavy object in the snow. Following these drag marks a short distance, I discovered what had been dragged. It was the hind leg of a moose with much meat and hide still attached. In following the drag marks in the other direction, I found the snow more and more tracked up by the foxes, wolves, and wolverines, and presently I arrived at the source of much happiness to a host of animal life — to all except the victim, a cow moose. Her end had come, as it must to all creatures.

The carcass was much eaten and the snow around it packed hard. On three or four knolls within a hundred yards of the remains I saw that foxes had curled up to await their turns while wolves and wolverine were finishing.

As I looked over the premises, it was soon evident what had happened to the cow. Tracks in the snow out among the spruces showed that, for a few days at least, she had been alternately feeding and resting. Tracks led from one clump of willows to another, and there were several deep, oval depressions in the snow where the moose had been lying down. When convenient, she had walked close to the spruce trees, where the snow was less deep and softer (the snow was about three feet deep). Did this taking advantage of the easier walking indicate a weakness in the cow? Not necessarily.

Then the wolves had discovered her. There apparently had been at least four wolves, for I saw tracks of four leave the area together. When the cow saw the wolves, she took a stand with five or six spruce trees behind her. Before her was an open area. It was probably routine for these wolves to test the moose they encountered in their travels. The development of the situation, I do not know. As the wolves dashed in from the sides or from behind, perhaps she was slow in defending herself. Perhaps she soon tired and the wolves were able to attack and injure her from the rear before she could turn and strike. Possibly these were especially hungry wolves and unusually persistent in their attack.

The evidence indicated that the cow had defended herself valiantly. In her maneuvering to fend off the wolves she had backed into and broken the lower branches of the spruce trees to a height of six or seven feet. Twenty of the limbs that had been broken off were one to one and a half inches in diameter at the break. After the limbs had fallen to the ground they had been trampled underfoot until they had lost their branchlets, which in turn had lost their twigs, and the twigs their needles. In two spots the lost blood had melted through packed snow, and under the carcass the snow showed a further loss of blood. There had been a prolonged battle, and this time the cow moose had gone under. She was an old animal.

10. Wolverine Trails

ONE WINTER DAY many years ago my brother and I were mushing our dog teams up a narrow creek in northern Alaska. A region seldom visited, it was many days from the nearest habitation. The spruce trees bordering the creek were almost the last; over that high range ahead of us there would be no more. We were proceeding slowly up the winding creek when we saw a lone track on the smooth snow. We stopped the dogs, ready enough to halt, and snowshoed ahead to examine the trail. There were the telltale five toes, the broad rounded track, and the pattern, and we knew it was a wolverine's.

We noted that this wolverine had been loping down the middle of the creek and had come to a sudden stop. The mingled scents of humans, dogs, dried fish, and babies (we were traveling with an Eskimo family) from our entourage had floated to him, a bend or two ahead of us. I doubt that he had experienced such smells before, but he showed no curiosity. He had fled into the spruce woods bordering the creek.

Mushing dogs, with the dog fights, the trail breaking, and the wondering if the dog feed will last, is not without excitement, but what we remember about that arctic day is that trail on the creek bed. We had met a wolverine!

Since that day I have followed many trails, but I have

gone for years without seeing a wolverine. Wolverines are seldom observed. Trappers long in the hills rarely meet them. Ernest Thompson Seton, the great naturalist, reported seeing only two during his lifetime. R. M. Anderson, over a long period of years devoted to travel and study in the North, saw sixteen, and most men of the North can readily recollect the few occasions when they have encountered one. The explanation lies partly in the relative scarcity of the animals. Although widely distributed, they usually are not abundant. Furthermore, they are wild and apprehensive, and they waste no time in making a safe retreat when they become aware of a peril. No doubt the wolverine has an excellent nose, and, judging from my meetings with him, his vision is not lacking in acuteness. His keen senses, wildness, and alertness largely account for his obscure existence.

The wolverine's wildness is illustrated by the following incident. About six o'clock one evening, as I was driving down the highway in Mount McKinley National Park, I saw an animal galloping away from the road as though terrified. I recognized it as a wolverine, although it seemed unusually slender, probably owing to shedding. It continued its headlong flight until far up the slope, where it sat up for a momentary look and then disappeared into a growth of willow. Its flight was exceedingly direct and strenuous.

On another occasion we saw a wolverine gallop up a long talus slope, again wildly hurrying from us. This one looked very much like a marmot, and I suspect that a marmot could easily be mistaken for a wolverine. But when he turned his head to one side to look back, as he frequently did, his long, flexible neck was obvious. He continued his flight to the top of the ridge, where he went out of view.

In appearance the wolverine resembles a miniature bear.

Heavily muscled, strong, and agile, it weighs twenty pounds or more. Charles Sheldon took one on the Toklat River in March, 1908, which measured a little over forty-three inches in length, including the tail, and about fifteen inches high at the shoulders and weighed thirty-six pounds. The jaws are well developed and articulated in a deep groove in such a way that they cannot be pried from the skull. This close articulation, together with the heavy musculature, suggests that the jaws are capable of administering a powerful, crushing bite. The color of the fur is blackish brown, the head and tail being lighter than the body proper, except for the two broad, tan stripes which pass from the neck, back along the sides, to meet at the tail. It is much used for trimming parka hoods because of its attractiveness, and because it is supposed to gather less frost from the breath than do other furs, such as that of the wolf and the dog, which are also used for trimming hoods. It so happened that the wolverine fur trimming my own parka hood collected considerable frost, but possibly a wolf or dog trim would have collected even more.

In April, 1949, Red Woolford and his partner trapped two male wolverines near Broad Pass about half a mile apart. One of the animals was hog-tied and brought in alive, not without considerable effort. When I saw him, he was inside a dark shed, fastened to a chain. Because it was feared that a collar would slip over his rounded head, owing to his heavy neck muscles, instead of a collar a harness arrangement was made that circled his back and neck in such a way that it held him securely; it was on the order of a lap-dog harness but different in details. When we entered the shed, we were met by growls so deep and hoarse that I wondered if he was securely chained. As my eyes became accustomed to the dark, I saw this fierce animal facing us in a threatening attitude, not at all daunted but ready to attack. Red ran the chain through a long metal pipe, which enabled him to lead him outside and hold him off at a safe distance from us.

There was almost a continuous struggle as the wolverine tugged and fought. Of special interest were his rolling tactics. He kept rolling over on his back, and sometimes he rolled over and over so that the chain became twisted. Such behavior in a trap might possibly help pry it out of shape or loosen a foot. Red stated that once, when his hound rushed in to attack the wolverine, the latter assumed a position on his back as though ready to rip with claws and teeth, and he succeeded in grabbing the hound's nose. His readiness to roll over on his back suggests that it is a common defensive attitude. Possibly he would face a wolf in such a position.

When the wolverine was returned to the shed and proffered water, he drank freely, lapping like a dog. It has been stated that wolverines suck the water in like a cow, and perhaps they do at times. In the old books this animal has been called the glutton, and there are many stories in print attesting to its extravagant appetite. I asked about the appetite and was told that it was not especially large, that the amount eaten seemed about normal for the size of the animal. No doubt a hungry wolverine, like a hungry dog, will gorge itself, but perhaps, if food is regularly available, the quantities eaten will not be exorbitant.

Tracks are as much as most travelers see of the wolverine, especially tracks in the snow. The wolverine has indeed a kinship with winter. It is then that his trails enhance the snowy landscape and add a rare quality to a region.

There are many other animals in Mount McKinley National Park whose tracks we see in winter. The moose plows deep furrows, the weasel leaves an erratic line of jump marks, the ptarmigan forms a lacework of foot prints and wing and tail marks. All have their characteristics, although under some conditions these may be confusing.

But we are mainly concerned with the tracks of the wolverine. Like other animals it has several gaits, and, to my knowledge, little has been recorded about them.

The wolverine's muscular legs terminate in big feet with well-developed toes and claws. Therefore, the track is large, similar in size to that of the wolf, for which it is sometimes mistaken. Under favorable conditions the five toes and the claws show, but on hard snow the inner toe may occasionally fail to register. The imprint of a front foot is much longer than that of the hind, because of an extra posterior pad, but the widths are about the same. Tracks of the front feet have measured five and three eighths to six and one half inches long, and four to five and one half inches wide; hind-foot imprints, three to four inches long and four to five and one half inches wide. In the same series of tracks there is some variation in length due to sliding and depth of impression, and in width because the spread of the toes increases with the speed of the animal. Anteriorly, the track is broad and rounded, much like a cat track. In loose snow the details are in varying degrees lost and the scuffing enlarges the tracks. But where the details are lost, identification can usually be deduced from the pattern of the four feet.

The wolverine has several gaits, each of which results in a characteristic pattern. For convenience we might list them as walking, trotting, loping (three-track pattern), loping (two-track pattern), galloping on hard surface, galloping in deep snow. In the running gaits the position of the hind feet in respect to the front feet varies. In the two-track lope and the gallop in deep snow the hind feet fall in the front-foot tracks.

In the three-track lope the hind feet go farther forward so that only one of them falls in a front foot track; in the gallop on a hard surface, the hind feet may fall far in front of the front feet.

Walking and trotting gaits of the wolverine may be considered together, because they have the same general pattern. The hind feet step in the front-foot tracks, leaving a trail like that of a man walking. The distance between the tracks measures from three inches to twelve or more. It is assumed that the shorter step is made when the animal is walking and the longer when it trots. The breadth of the trail is seven or eight inches. If the snow is deep, the feet show drag marks and there is sometimes a trough effect.

The lope (three-track pattern) is the most common gait. In this gait the hind feet are brought farther forward than in the two-track lope. Usually, the posterior hind-foot track partially covers the anterior front-foot track, and the other hind foot falls a few inches ahead of this track. This gait results in three tracks for each jump. The three tracks characteristically fall in an oblique line, so that the trail is a succession of oblique lines which can be recognized from afar. These lines may point to the right or left, but usually they are rather consistently one way or the other for a considerable distance. The oblique pattern is due to the hind legs' being brought forward a little to one side. If the hind feet come forward on the left, the posterior right-hind track joins the anterior front-foot track, and the left-hind-foot track is up ahead by itself. Movies of a wolverine coming toward me show both hind feet to one side of the body when the hind quarters are off the ground.

The lengths in the row of three tracks have been measured from twenty-nine to forty-seven inches, the longer measurement representing an animal traveling fast. The distance between one group of tracks and the next ahead vary from seven to nineteen inches. The three-track lope is the cruising gait, the one most commonly used. It is a bouncing gait, sometimes referred to as a "humping" along. A trackwalker on the railroad said he had seen a wolverine which appeared to jump with all four feet in the air at once (like a mule deer); that is sometimes the impression one gets.

A two-track pattern lope occurs occasionally when the hind feet fall in the tracks of the front feet; two tracks result from each jump, one a little ahead of the other. This pattern is similar to that usually found in the mink and marten. I have noted this gait where the animal was going up a slope or sinking three or four inches in snow; in other words, where travel was somewhat laborious.

The gallop is the speediest gait, all four tracks showing separately, and might be called the four-four track. The hind feet alight far in front of the front-foot tracks; and the distance between jumps becomes considerable. The four tracks may have a spread of sixty-seven inches, and over eight feet may be covered in a jump. A wolverine fleeing from a wolf registered the longest jumps.

In deep snow the trail of a galloping wolverine is a line of holes, each hole ten or twelve inches wide. For each jump there is one hole, hence we might call it the one-one gallop. The forefeet strike side by side, and the hind feet alight in the forefoot tracks. The snow is usually so loose that details are lost. Between the line of single holes there usually are double drag marks made by the feet. Sinking six inches, one trail showed a maximum of forty-five inches between jumps; another, sinking six inches, showed eighteen to twenty inches between jumps. A wolverine running from a wolf took this gait when it encountered soft snow.

Glimpses of the wolverine are among a naturalist's most notable experiences. The wolverine is a steady boarder at any carcass he discovers, and if one can find one of these there is a good chance to watch him dine. Some carrion freezes solid before it is found; then it is slow work, even for a wolverine, to gnaw off a meal, and he spends considerable time there.

On the morning of March 14, 1950, while driving out in a snow jeep, I observed from a distance a dark object beside a dead moose, and, looking through the field glasses, I recognized a wolverine. He was feeding on the shoulder, changing his position vigorously and frequently, for better gnawing. Once he lay on his side while working down deep in the chest. At short intervals he threw his head up for a look around. After I had watched for some time, I drove to within half a mile of the carcass. When I stopped, I noticed that the wolverine was leaving, but he gave no sign that the snow jeep had scared him. He loped along at a cruising gait without looking my way and traveled up a small tributary just under the bordering snow drifts. A number of trails leading up and down the creek bottom showed that the wolverine had made many visits to the dead moose. These trails frequently led to lone spruce trees which the wolverine had sometimes climbed. There were claw marks on the bark, and bark fragments were scattered on the snow beneath the trees. One tree having considerable slant and therefore easily climbed had much of the outer bark worn off and was apparently a favorite for climbing. Perhaps the wolverine found this tree useful as a lookout. I surmise that the wolverines use the trees as rubbing posts, for they seem to rub themselves frequently. In many places the animals had rolled in the snow, and in two or three places a shallow hole had been dug and a dropping left and covered over with snow.

On March 28 four inches of new snow lay on the ground, covering the accumulated trails and furnishing a clean surface for fresh ones. As I came opposite the moose carcass, I again saw a dark object on it. I thought it might be the black wolf, as it sometimes was, but it was a wolverine. At first, all feeding on the moose had been done in the region of the shoulder, but now a shaft had been sunk into the hindquarter near the tail. The warm March sun, shining on the dark surface, had softened the tail region a little, and the wolverine or the wolf had worked through the hide. The wolverine, as usual, was working vigorously at the carcass, changing its attack back and forth between the shoulder opening and the tail region. His movements were quick and sudden. At short intervals he stopped feeding to look around, and twice climbed up on the moose for a better view. The object of its watchfulness was probably the wolf. A red fox, curled up on a drift fifty paces away, slept on as though he had not a worry. He was awaiting his turn and making use of the time by sleeping. He probably kept an ear cocked toward the wolverine, and if the wolf showed up, he need not worry about being able to escape him either.

I drove the snow jeep slowly to within a half mile, without frightening anything. The wolverine continued its feeding, and the fox remained asleep in his bed. A raven lit in a willow bush beside the dead moose, giving the wolverine such a start that he galloped fifteen or twenty yards away. But he returned at once, stopping briefly to rub his sides on the snow. The raven remained perched in the willow. After I had watched from this new position for a time, the wolverine departed, alternately walking and loping, using one gait about as much as the other. His short, bushy tail drooped, and his head was held low. He stopped to sprawl and roll, rubbing himself on the snow over an area a dozen feet across. He also rubbed against some slender willow limbs, causing them to bob up and down and wave in the air. Then he moved a short distance to an open slope and lay down. (His bed proved to be twenty-two inches across and of an irregular shape.) His watchfulness continued, for he rose up occasionally to look around. When I moved forward with the snow jeep, he galloped away, jumping thirty-seven to thirty-eight inches in the soft snow. Tracks showed that, in approaching the dead moose to feed, the wolverine had circled so as to be downwind, probably to learn if anyone was there.

On August 15, while driving in a car, I came suddenly upon three wolverines in the road. I wanted a picture, but the camera was put away. If I stopped the car, the wolverines would no doubt leave the road and disappear in the rocks and brush, either up or down the steep slope, so I followed slowly. All three galloped down the road ahead of the car, turning their heads on their flexible necks to look at us as they ran. After a short chase I stopped to set up the movie camera. One of them went up the slope, two disappeared around a curve, and I didn't see any of them again. I suspect that these three wolverines consisted of a mother and two offspring.

An interesting observation was made one year on July 21 by Mr. and Mrs. Edwin C. Park along Igloo Creek. For over two hours they watched a mother and two young almost as large as the mother, and observed her nursing twice. The breeding habits of the wolverine are reported to be similar to some of the other mustelids, in that the development of the embryo is arrested at an early stage and remains dormant for several months before development begins again and implantation occurs. The breeding apparently takes place in the summer, but the young are not born until some nine months later. It seems surprising, in the foregoing incident, that such large youngsters should still be nursing, yet perhaps this is less astonishing than grizzly cubs nursing into their third summer abroad.

While eating lunch on May 22 George Stiles and I were watching a group of ewes and lambs as they moved downward toward a little creek at the base of some cliffs. After our lunch we stalked them for a movie, gaining a point a little above the creek and just across a draw from them. But they took fright and hurried away without presenting any picture opportunities. We were sitting on the point to catch our breath when up the creek I noticed a dark spot on the snow coming our way. To my surprise, it was a wolverine — bounding along at an easy gallop on the snow-covered creek, coming steadily and purposefully as though to meet some appointment. The gentle slope of the creek bed made travel easier, and he was making good speed, though his gait seemed effortless. At each jump the hind legs were thrown high and to one side. Coming nearer, he was hidden by the bank, but directly below us he came out of the creek bottom. He glanced toward us but continued on his way through the low bushes without stopping. The incident was so unexpected it seemed like a dream and left us exclaiming in hushed voices.

Tradition says that the wolverine has no fear of other animals. According to Hearn, as quoted by Seton in his *Lives of Game Animals,* a wolverine was known to drive a wolf away from a deer it had just killed. A mountaineer friend of Seton's told of two wolverines driving a black bear from the remains of an elk. R. M. Anderson saw three dogs, "including a famous bear-dog," attack a wolverine; they found it so fierce that they finally gave up and let the quarry go its way. Fry, as quoted by Seton, reports two black bears giving way to a wolverine at a cow carcass, and three coyotes leaving a dead horse on the approach of a wolverine. The same observer reports that two cougars left deer remains when a wolverine approached. Charles Sheldon writes in *The Wilderness of Denali* of having seen a lynx run from a dead sheep and a wolverine at the same time approach it. Incidents such as these do not necessarily mean that the wolverine could whip these larger animals. More than likely the retreats were made to avoid annoyance rather than from any fear of being overcome. Nevertheless, the wolverine appeared to be respected as being able to fight back.

At Mount McKinley National Park tracks revealed on three occasions that a wolf had chased a wolverine. That the wolf gave chase does not mean that he would close in if he overtook the wolverine, for the chase may have been a sporting event. However, a pack of wolves might cause a wolverine some trouble.

The first incident took place on December 2 and was interpreted from the tracks. While a wolverine was loping along a snow-jeep trail at an easy gait, a wolf was moving slowly up from a creek bottom toward the same trail. Their ways met. The wolverine must have scented the approaching wolf, for it halted and then reversed its direction. Upon first turning back, the wolverine was not much alarmed, judging from the slow, loping gait, but after he had traveled on the back track for about twenty yards, he must either have seen the wolf or gotten a warmer scent, for he broke into his fast-

est gallop. After traveling for 150 yards, he left the road and
galloped through the soft snow, sinking deep at every jump.
The wolf, on coming to the road, had followed the wolverine
at a fast trot, but after a chase of about 350 yards it had apparently lost interest and resumed its lone way. The wolverine had done his best to remove himself from the vicinity
of the wolf; and the wolf had not seemed very anxious to
overtake this formidable fighter.

Tracks observed on March 29 at the carcass of a moose
showed that a lone wolf had rushed toward a wolverine that
was lying on a snowdrift beside a leaning spruce. This tree
was apparently a refuge, and the wolverine had been resting
beside it for that reason. In this emergency, he climbed to
safety and peace up through the thick, brushy branches. The
litter of loosened bark and broken twigs lying on the snow
beneath marked his progress up the tree and showed that he
had climbed to the very top. There were scratch marks on
the trunk, and a few wolverine hairs clinging to some of the
sharp knots. On this occasion I should guess the wolf was
motivated by an exuberance of spirit.

On another occasion I was following the trotting tracks of
a wolf going over hard, drifted snow near the top of a bench
bordering a small tributary of Jenny Creek. He had a good
view of the creek bottom, the willow thickets, and the few
lone spruces. From his vantage point he had seen a wolverine coming slowly up the creek. Here was another opportunity for the wolf to indulge in a diversion, and he was soon
in full gallop. The wolverine had seen the danger and had
also broken into a gallop. His course was directly up the
creek toward a spruce; the wolf ran almost parallel to the
creek but slanted toward the wolverine's line of travel, as

though he, too, were headed for the spruce tree. On the hard snow, the wolf's widely spaced tracks showed great speed, but when he reached the soft snow in the bottom, he broke through and his trail was a series of deep holes, where he had wallowed. The quarry could not make speed in this snow either, but he managed to reach the tree ahead of the wolf. The snow around the spruce was marked with many wolf tracks. There were several spots where the wolf had sat on his haunches, probably looking up at the wolverine. The wolf had trotted to the top of a low ridge for a look around, then returned to worry the wolverine some more. Eventually, the wolf tired of his sport, and the wolverine was able to continue on his journey.

The cache, typically a miniature cabin on stilts, is characteristic of the north. It is a structure that is both picturesque and useful. When a trapper builds a cabin, he also expects to erect a cache on four poles, because his provisions, even when left in a cabin, are not safe from disturbance. The animal most closely associated with the cache in the minds of many is the wolverine. (It probably should be the bear!) The wolverine is known widely for his habit of misappropriating goods; because of the remarkable ability he has exhibited in breaking into stored provisions, he has been credited with superhuman strength, a reputation at least partially deserved. But his misdemeanors have helped to foster the pleasant architecture of the cache which has such universal appeal. Because of his climbing ability and that of the mice and others, the poles are ornamented with tin or stovepipe, which is wrapped around a section of each pole. The smooth surface presents no hold for the claws and is an effectual block to climbing higher and gaining access to the stores. I like to think of the cache as a gentle way of settling some of the differences between ourselves and the other animals in the hills, so that we can live amicably together.

The wolverine has his own caches to enjoy and worry about. It is not easy to be secretive about them, because of the many sharp eyes and noses in the woods. If a bone or other part of a carcass is carried away, it leaves a give-away scent on every twig and grass stem it touches along the trail. Consequently, when someone like the fox, who follows tracks of all kinds, learns from these scents what the wolverine is up to, he follows the trail and sooner or later finds the cache and appropriates what it contains — unless the wolverine is there. This works both ways, for, the next time, the wolverine may discover one of the secrets of the fox. Thus we find that the caches, regardless of who makes them, tend to serve the community.

A moose carcass north of Six-Mile, quickly dismembered because it was fed upon by wolves and others while still fresh and unfrozen, was frequently attended by a wolverine, who made many trips, carrying away booty. He was providing for the near future, when this food source would be gone. (The wolverine that fed at the frozen carcass did no caching, because the meat could be only slowly gnawed loose, and no surplus for caching was readily available.)

On one occasion a wolverine stole a frozen section of a moose leg which Ranger Bill Nancarrow and I had hung in a spruce by a rope. I found the drag marks of the leg, and wolverine tracks to one side. The trail led through the woods to the creek bed, where I assumed I would find the cached leg, for across the creek was a steep ridge. The leg weighed perhaps thirty-five or forty pounds, and I marveled that it had been dragged so far. Then, to my further astonishment, I found that the trail led up the ridge, so steeply that my snowshoes took an occasional slide with me. The wolverine had done much tugging to get up the slope (some two or three hundred yards), especially in places where the leg had caught in willows. After gaining the top of the ridge, the wolverine had continued over rolling terrain and then gone down into a ravine — where I turned back. A fox had followed the trail ahead of me!

About three weeks later, far up this same ravine, I noted fresh digging in the snow on a steep slope and found a hard-packed platform on which reposed the bones of the moose leg, still articulated but cleaned of all hide and flesh. The end of the humerus had been chewed and hollowed out as though the marrow had been sought. While examining the bone, I thought I heard a sound like the crackling of branches but suspected it was my parka rubbing on twigs. More sounds called my attention to the top of a spruce; there I saw movements and a patch of brown and yellowish fur, and I recognized a wolverine. There was a throaty growl, then several growls. When I maneuvered away from the tree for a picture, the wolverine scrambled down the tree and plunged away through the deep snow, hidden at once by a heavy growth of alders. He probably knew about other cleaned moose bones scattered widely over his domain.

Next winter the wolverine will again leave his snow trails in McKinley Park. On the far hillside there will be a trail with a three-three pattern, and you will know that he has passed that way. You will meet his track in the valley and on the crests of the ridges. Here, at least, we may hope that his trails will lead far into the future.

11. Reynard of the North

THE RED FOX in McKinley Park is numerous and prosperous. He comes in many colors: red, black, silver, and smoky, with intermediate variations. As many as three distinct types may occur in the same family of pups. The feet are black, and the tail is tipped with a white blossom of varying extent. Some foxes have slight special marks which identify the individual. The white tail tip may be especially long; one pup had a white patch on its black foot. Split-ear, discussed in the next chapter, had a deep split in the ear. Some may be recognized by their general behavior. They are, of course, all individuals, each with his own special character.

Some of the foxes which I could recognize, I saw over a period of months or even for two or three years. Data secured from well-marked foxes showed that individuals tended to spend their lives in a restricted home range. One fox that I saw many times was a male that I first observed on April 24, 1939, near a cabin on Igloo Creek. Because of a missing lower canine tooth, he was identifiable when close by. During the summer he was rarely seen, but in September and October he often came to the cabin and grew quite tame. He learned to come when I whistled, and once he followed me a mile from the cabin when coaxed with tidbits. In 1940 I lived elsewhere and had less opportunity to watch this fox, but I did see him twice, on June 21 and November 17. While I stopped a few days at Igloo Creek in February, 1941, the fox appeared, looking for scraps. During the summer of 1941 he was seen regularly in the same vicinity. This fox, then, lived in the neighborhood continuously for at least two years and three months. No information on the extent of his range was secured; when seen, he was always within a few hundred yards of the cabin.

At Sable Pass a pair of unusually colored cross foxes used the same den in two successive years. The same foxes were seen in the area in winter.

Near the cabin on Toklat River a cross fox lived for at least three or four years. The building was vacant most of the year, but whenever it was occupied, summer or winter, the tame fox made its appearance for scraps. It was not fed regularly enough to restrict its movements.

An unusually beautiful silver fox lived on the east side of Sable Pass. I first saw him on April 23, 1939. He discovered me when I was only seven yards away, looking at me with eyes which, in the bright sunlight, were only slits, then continuing unhurriedly on his way. I was often close to him, but this was one of the few times he ever deigned to look directly at me. Usually he went about his affairs as though I did not exist, even when I was near him. He was always seen in an area about three miles across, south of the draw where I suspected he was denning. In 1939 I saw him during the summer and again in late October, shortly before I left the park. In the summer of 1940 I continued to see him at frequent intervals. In November I saw him on two successive days near the same spot. In the spring and summer of 1941 I saw him frequently. Twice he was carrying food, apparently to his den. This fox was known to occupy a definite range over a period of two years and three months. Since this fox lived some distance from any cabin and was never fed any scraps, he was existing in an entirely natural environment.

Although some foxes seem to live all year in a restricted home range, there may be, at times, general movements of foxes. A trapper told me that during a mouse epidemic on his trapping ground there seemed to be an influx of foxes from other areas. This was correlated with a scarcity of foxes reported by other trappers some distance away, where mice were scarce. The increase may, of course, have been due to a good fox crop as a result of a large carry-over of breeders and an abundant food supply.

Little was learned concerning the dispersal of the young each year. They probably wander about considerably and fill in unoccupied territories.

The proximity of some of the occupied dens to one another suggests the size of the home ranges, but, of course, there may be a deep overlap of territories. Two dens were five miles apart. Two others were six miles apart. About halfway between the latter two, I was sure there was a third den, which would place these dens about three miles apart. Two dens were only about one mile apart.

I found dens in various situations — in the open and in the woods, on sunny knolls far up the slopes and on the flats. Most of them were dug in sandy loam, but a few were in hard clay. Generally, they were found on south-facing knolls, where the soil was somewhat loose. A typical den has from six or seven to nineteen or more entrances.

I found ten occupied dens from 1939 to 1941 without making a special search for them. In 1940 I discovered five along a thirty-two-mile stretch of the highway. In this same stretch the approximate location of four other dens was known. Clearly, there was an abundance of foxes in the park.

During the summer the foxes sometimes changed dens; in two instances changes were made when the foxes were undisturbed, so far as I knew. There usually are several vacant dens near an occupied one. The same dens are sometimes used in successive years, as is the case among the wolves. Split-ear, a female, had the same den two successive years, but had denned two miles away previously.

The foxes in the park exist in a rather ideal fox world. Man here is their friend. And, normally, there is an abundance and variety of food. When mice are abundant the foxes can subsist royally on their mouse hunting; when the snowshoe rabbit is available, they hunt them efficiently; and ptarmigan, always present in sufficient numbers to furnish an occasional meal, may serve as a dependable basic food when abundant. The hibernating ground squirrel, in safe coma status all winter, becomes a major item in the summer diet. To these important food sources we may add carrion (moose, sheep, and caribou), which is highly welcome in winter. Supplementing the animal foods, the foxes use extensively the wild orchard that is many miles in extent. They are especially fond of the blueberries, which they eat in quantity in summer and to some extent in winter. And the abundant black crowberries are also on the menu. I studied the food habits of the foxes statistically by gathering upward of a thousand scats for analysis and devoted considerable time to watching their hunting activities.

During the winter and in the absence of rabbits, mice usually form the staple diet. Meadow mice are easily captured even when their activity is carried on under moderate depths of snow. The fox locates the mouse by scent or hearing, then pounces so as to break through the snow directly over the mouse and pin it down. Even if the first pounce fails, it may block runways so that escape is shut off and the mouse is captured in the ensuing scramble.

Ground squirrels are abundant and much used. They hibernate in winter but are obtainable from April to October. Their remains were found in 307 of the 662 summer scats.

On June 17, 1941, I saw a silver fox moving steathily along a ditch beside a road, hunting for squirrels. At intervals he stopped to look. Meanwhile, the squirrels from surrounding points were calling sharply and disappearing into their holes as he neared them. After moving along slowly with no success, the fox made a quick dash of two hundred yards through some hummocks, apparently to get into new territory where his presence was not widely advertised. After making this long spurt, he stood still, watching, then moved out of my view among the hummocks. When hunting mice, the fox makes no attempt to hide, but while hunting ground squirrels, he keeps hidden as much as possible.

On July 2, 1941, I again observed this same silver fox hunting. He captured three mice in ten or fifteen minutes. I had to leave then, but when I returned an hour later, the fox was carrying a ground squirrel. Three times he laid it down in order to pounce for mice. Once he lay watching a hummock for several minutes, waiting for a mouse to come out. Then, weary of waiting, he picked up his ground squirrel and trotted homeward.

A fox appeared at the Mount Eielson tent camp and quickly captured three of the ground squirrels which had been tamed by feeding (a good example of one effect of domestication). After catching the three squirrels, he piled them so that the second squirrel crossed the first, and the third crossed the second. Then he seized all three where they crossed each other and left for his den with the heavy burden.

On October 14, 1939, a red fox appeared at our cabin looking for food scraps. He was about to feed on some morsels we had tossed to him when he suddenly turned, crouched, ran forward a few steps, and pounced. We did not know what he had pounced upon until he lifted his catch — our pet ground squirrel, which had not been out of its burrow for several days and this day was probably taking its last look around before the long winter's sleep. My assistant and I looked at each other open-mouthed, and then he said, "Well, that's nature."

The fox with this treasure would not trust us, and he galloped away to feed in security.

Once a fox chased a ground squirrel into a hole and spent considerable time digging for it. Another fox ran for fifteen yards and pounced at a ground squirrel but missed him. He put so much energy into his pounce that he rolled over. As this fox continued on his way, the squirrels in the neighborhood sat on their hind legs and scolded him loudly, and sparrows darted at him.

Blueberries are common over the fox range and may at times become a highly important winter food. They began to appear in the droppings as early as September 13, and the foxes fed on them throughout the winter and into April. Mice during this time were present in moderate numbers. When foxes were first noted feeding on this berry, mice were available in sufficient quantity to furnish the foxes ample food. One infers that they are especially fond of blueberries. But no doubt the blueberries are at times eaten from necessity. Lee Swisher said that in the winter of 1939-1940, when mice were scarce, foxes were living almost entirely on blueberries on his trapping range north of the park. They dug down through the snow for the berries lying on the ground. That winter, probably because of the scarcity of mice, foxes captured in traps were eaten more often than usual by other foxes.

The fox in his northern world comes in contact with many species. I do not know of any significant relationship between the fox and the grizzly bear. Both feed considerably on ground squirrels and blueberries, but there is enough of this food for all. On June 28, 1940, a grizzly digging out ground squirrels was closely followed by a fox. While the grizzly excavated, the fox lay on the grass nearby. Sometimes the grizzly followed the fox, which retreated slowly before it. The fox remained near the bear for the hour and a half that I watched them. Knowing that the fox had a den somewhere in the vicinity, I wondered if it was trying to lead the bear away. Shortly before I left, the fox was sitting some distance up the hill, tall and straight, watching the bear below it. Then it trotted over a knoll out of sight.

There was no evidence that the eagle affects the fox population. Foxes are so numerous and spend so much time traveling in the open, treeless areas during daylight that the eagle would have many opportunities to prey upon them if it were able and so inclined. The lack of any remains in any of the 774 eagle pellets I gathered indicates that foxes are rarely eaten by the eagle. There is a possibility that young eagles sometimes attack foxes with serious intentions, but they probably learn that it is a dangerous venture.

Eagles have been observed swooping at foxes just as they swoop at almost every other mammal in the park, including grizzlies. Many of these maneuvers are in sport. On May 14, 1939, I saw an eagle soaring thirty or forty feet over a fox which stood in the open looking grim and tense, his tail straight up in the air. (Foxes often assume this pose when excited. I have seen one holding his tail in this manner after pouncing on a mouse which he still searched for, and I have seen pups at a den take this stand.) When the eagle saw us, it flew away, and the fox relaxed and trotted slowly over the tundra. The pose taken by the fox is apparently one of readiness to ward off an attack. If the fox should run, it would give the eagle an opportunity to strike.

J. Dixon describes an incident in which he saw an eagle swoop at a fox which was crouching in the open. A second fox was driven out of a culvert nearby, and when he galloped away, the eagle attacked him as he ran, but the fox avoided its swoop and went into a cleft in a rock. The fox possibly would have behaved differently — probably would not have run — if he were not escaping from humans as well. To avoid humans he must run; to avoid an eagle he must stand ready for attack or discreetly retire to cover. The same maneuver was not suitable for both enemies. The incident must be interpreted in the light of this knowledge. Its significance, as I see it, lies primarily in showing that the fox can avoid the stoop of an eagle.

On June 7, 1941, I saw an eagle standing beside the entrance of one of the burrows of a fox den. I am not sure just what was taking place. The eagle would reach into the entrance with its beak and then withdraw as the fox's head emerged from the hole, its jaws wide open and snapping. When the fox's head disappeared, the eagle would stoop over the hole, only to draw back quickly as the fox's open jaws appeared again. This was repeated four or five times before the eagle flew away and the fox came out to lie on the grass. Possibly the eagle had first been attracted to the den by the presence of a dead ground squirrel or some other scrap of food.

Foxes seem to have no fear of wolves. On July 23, 1940, a red fox sat watching a wolf sixty yards below it. Later the fox trotted along parallel with the wolf as the latter traveled across the slope. When the wolf descended the slope, the fox followed it a short distance down. The actions of the fox showed a confidence in its ability to evade the wolf.

On July 19, 1941, some members of a road crew saw a black wolf sniffing about the vicinity of a fox den. An adult fox followed the wolf closely and barked at it from a distance of a few feet. Once the fox ran off as though it were trying to entice the wolf away, but returned when the wolf did not follow. The wolf paid no attention to the fox. It was searching for cached food items. Wolf scats at the fox den showed that the wolves had visited the den previously. Mr. Brown of the Alaska Road Commission told me about the incident, but though I hurried to the den, the wolf had gone before I arrived. However, I saw the wolf about two miles beyond the den and later saw it catch a calf caribou and photographed it as it fed and then cached the remains. The incident again illustrated the fact that foxes have full confidence in their ability to run away from or avoid a wolf. I have observed magpies and short-billed gulls searching for morsels at a fox den, a source of food which many animals know about.

At Teklanika Forks, in 1939, I observed a gray wolf sniffing about a fox den, perhaps looking for food. Wolves probably visit many fox dens in search of scraps, especially if their food supply is a little scant.

The relationship between the wolf and the fox seems to be

one of mutual gain. The wolves benefit by having available a large number of old fox dens which they can enlarge for their own use. It is a simple matter for the wolf to enlarge a burrow and much easier than digging a new one. Although the fox burrows are too small for the adult wolves, the pups can use the entire system of burrows. So far as I know, all the wolf dens found in Mount McKinley National Park were renovated fox dens.

Although the fox loses a few food items when a wolf ransacks its den site, the loss is insignificant. If there is food present, it is a surplus which can be spared, whereas, if food is scarce, there will be none lying around.

Generally, when a wolf makes a kill of a large animal, this is shared by the fox, for, after the wolf has eaten, there is usually some of the carcass remaining. Signs at a great many of the carcasses examined in the field showed that foxes had shared in the spoils. Much of the food supply is made available to the foxes in winter when their food is scarcest. Here the fox's large gain is the wolf's small loss. Since the fox eats much less than the wolf, the loss is usually not serious and to a degree represents a surplus, although at times, of course, this surplus may later be needed by the wolf.

Wolves often cache the remains of a carcass after eating their fill, and the foxes commonly track down the wolves and rob the stores. Such an incident took place on October 4, 1939. A wolf had killed a lamb on Igloo Creek in the morning and, after feeding, had removed a part of the carcass and carried it away with him. The ground was covered with two or three inches of snow, sufficient for good tracking. The first indication I had that the wolf was carrying a load was the blood and hair on the snow where he had placed it on a knoll when he had stopped to look around. At two other little prominences also he had laid the meat on the snow. Although the wolf track was only a few hours old, a fox was ahead of me on the trail. He probably had gotten the scent of the sheep meat from the vegetation along the way. In one place the wolf had backtracked for fifteen yards, had jumped off the trail eight feet to one side, and then had wandered about in several loops. At this point the fox tracks circled about as though the fox had been having some difficulty in unraveling the trail. The wolf resumed his direction northward through some wet tundra, walking in shallow puddles of water, apparently by choice to destroy his scent. After passing through some woods he came to Igloo Creek, and there his trail disappeared. The fox had come to the stream and had stood with front feet on a snow-covered rock in the shallow water beside the shore, apparently sniffing for the scent. In two other places, tracks of the forefeet on a rock showed that the fox had stood facing the stream looking for the lost trail. The wolf had walked in the water for fifteen yards and had come out on the same side of the stream again. The fox and I both followed down the stream until we came to his tracks.

Down the bar three hundred yards the tracks led directly across the shallow stream. Here the fox, without hesitating, had also crossed the stream. After following the bar a little farther, the wolf went into the woods, where his trail made an S. And there beside a tree was the cache, already raided. I caught a glimpse of a cross fox carrying something, probably sheep meat. All that was left at the cache was much loose sheep hair. The cache had been covered with lichens and snow.

Beyond this point the wolf and fox tracks continued for 150 yards to a second cache beside a hummock. The wolf apparently did not believe in having all his eggs in one basket.

But both baskets had been robbed, for the second cache was also raided. The wolf track continued through the woods and led up a long mountain slope. Blueberries on the ground, which he squashed as he walked, colored many of his footprints purple. The fox tracks stopped at the second cache. The fox probably knew that the wolf had cached all his load. The behavior of the wolf seemed to show that he was aware that he would be followed by foxes, for it seemed he made deliberate attempts to throw them off his trail.

All the data gathered on the wolf-fox relationships strongly support the conclusion that the fox population has not been harmed by the presence of the wolves in Mount McKinley National Park and the adjacent region north of the park, and that both species can subsist in the same region in good numbers.

12. Split-ear, The Fox

OF THE MANY FOXES that I watched during my days in Mount McKinley National Park, I never knew any as intimately as I did Split-ear. When I saw her in June, 1956, she was not looking her best. No longer was her coat alive and vibrant as it must have been when she was frolicking over the snow with her mate in the sharp, below-zero nights of February and March. Those happy days had been followed by a scraping-out and inspection of the many dens she knew about, then by a heaviness and, finally, by the birth of five pups in a chamber along the bank of a shrubby ravine in the open tundra. Then followed three or four weeks of an underground nursing schedule which involved crawling in a crouch into one of the five burrows that led to the chamber where the young were huddled. The effects of her activity, condition, and the season were evident. Her long coat of reddish fur was shed over upper legs and hips, and the result was a skinny look that was not at all becoming.

At this time one wondered how she could have attracted a mate, especially one of the most beautiful foxes of the hundreds that roamed the tundra of McKinley Park. He was a silver-tipped black fox whose regal appearance was accentuated by conspicuous cream-colored ruffs. He was indeed handsome. But if one examined Split-ear's profile and forgot about her lean starved hips, it was obvious that she had truly classical fox features capable of causing any fox prince to bark his explosive, pheasantlike bark in her behalf, as a challenge to any approaching rival. In those glorious nights of spring, the ruffs and the profile, so pleasing to the human eye, were perhaps unimportant as two spirits sped across the wind-blown snow that covered their tundra.

The attachment between the vixen and her mate is not a short-lived affair that wanes to something commonplace following the period of the more intense passions. The vixen continues to show the most ecstatic behavior toward her mate, who devotes his time to long hours of hunting for her and the pups. She greets his return with extreme affection — tail wagging, wriggling, jumping, fawning, and face licking. His homecoming is always a happy occasion. I did not observe the male reciprocate with a show of fondness for the female, but perhaps this is not to be expected. First of all he is the "lordly male" following the male protocol of reserve and dignity. Then, too, he comes home weary after hours of hunting, and his whole being seeks rest and relaxation.

Split-ear's handsome mate I never saw during the day, and she apparently did not expect him until evening. But other males I have often seen resting at a den through the day. Perhaps my presence and his wariness were the answer. On a few occasions, in the evening, Split-ear became utterly frantic with impatience in awaiting his arrival. On July 6 I approached the den at six-ten P.M. The vixen, her coat somewhat wet and bedraggled by a recent shower, trotted to the den and touched noses with one of the pups. Then, for the following two hours and thirty-five minutes, she was restlessly occupied looking for her mate. Trotting or galloping, she hurried a hundred yards or so to prominences east of the den to watch; then, sure that her mate would come from the west, she dashed to a western lookout to gaze intently into the distance, eyes and ears alert. At times she curled up to sleep, as was her wont at the den, but in a few moments she would again be up and looking. When at last the male arrived — from the east — she was on the other side of the den. She spied him and galloped her fastest to greet him with the usual show of happy emotion. The male was undemonstrative and weary and soon curled up on a prominence, where he licked a paw and yawned widely a number of times.

And our impatient vixen! She picked up the ground squirrel the male had dropped before her and carried it to the pups at the den. Then, at a lope, she left, apparently to hunt. Was her impatience based on a desire to be off on her own excursion? This evening it may have been. Waiting for the male to bring food might also, under conditions of scarcity, be a cause of impatience. But on various occasions Split-ear, and other vixens that had food and did not dash off to hunt when the mate came home, would meet the home-coming male with exuberant affection. But emotions are no doubt mixed at times.

When I observed wolves, I found supplemental adults, both males and females, living amicably with a mated pair at a den. Some were at least two years old and so were not yearling offspring delaying departure from the parents. In spite of these observations, it was a surprise to discover something along the same order among foxes, which are not a pack animal like the wolf. The first inkling of this was on July 23, 1950, when I saw three foxes at a den. Two pairs of foxes have been 'found in a single den in the eastern United States in the midst of a heavy fox population, but this was a little different. Here, only the one litter of pups was present.

In 1955, on June 1, I saw a fox on a long slope approaching the den briskly, jaws loaded with mice and a ground squirrel. He detoured around me, and as he approached a knoll, he was met by his vixen, who jumped forth jerkily, wriggled, and swung her tail for joy. The male, intent on his mission, walked to the center of the den, where he deposited his load. The female ate a ground squirrel, cached a mouse, and galloped happily to the male, who was up the slope lying on his side. But he was too weary to play and walked away a few yards, while she crouched and jumped before him. When she approached him later, he made a token effort at play, but soon lay down. The way of it was that she was full of energy, having been baby sitting for days, and he was dead tired and wanted rest. He roused himself later to uncover a cached mouse and then draped himself over a hummock, thus relaxing a different set of muscles. Two hours after the male's return, a third adult fox, a male, that had been resting at the den all this time stood up and fed on a morsel. He walked up to the male and lay down close to him, wagging his tail — whacking the ground on the downward stroke. The male snarled at thus being disturbed, and the extra fox soon moved away and trotted off to hunt. He was treated with less deference than that accorded the female.

There were at least three supplementary adults at Split-ear's den in 1956. I saw from one to three on seven occasions. These nonparents, which functioned like maiden aunts or bachelor uncles, were treated passively by Split-ear and her mate — their coming and going meant little. When one made an appearance, it was noted and that was all. One evening, when Split-ear was impatiently awaiting the male, she met one of them and they stood together briefly and separated. The extra foxes brought food and called forth Split-ear's pups, played with them, and uncovered caches in the den area; they were right at home and behaved like parents. One evening, one of them baby sat while Split-ear hunted. It is not unlikely that these foxes were former offspring. Another possibility is that they had lost their own young and were satisfying their parental emotions at a neighboring den. One can only conjecture about their relationships.

At her den, Split-ear reacted to humans as she would to a moose or caribou, perhaps with even less concern. My near presence appeared not to alter her family life at all. On June 4, when I first observed the young, they were wobbly, chubby, blackish, and blue-eyed. Thick, short legs and chunky body gave them the look of little bears. As they grew, they became slimmer, the legs longer and more trim, the coats browner, and the eyes brown.

To call forth the pups, the mother walked to the mouth of a burrow and made a low, repeated murmuring sound, "mmm, mmm." If there was no response, she would repeat her message at another entrance. Usually the mother stood with eyes half closed while nursing; twice I saw her lie on her side, and occasionally, after the regular nursing, a pup or two might persist as she sat on her haunches. A session lasted two or three minutes and was generally terminated by the vixen stepping away. The pups crawled about the den and, when quite young, often retired of their own accord into the burrow. At this time mice, squirrels, and other meat items were also a part of their diet.

When Split-ear wished the pups to retire, she uttered a single short sound, "klung," that seemed to come from deep in her throat. Thus we have fox language: "mmm, mmm," come forth; and "klung," retire, go underground. Perhaps "klung" means danger too, for when the pups were older and heard this signal, they looked around as though to see for themselves why they should obey. The hoarse, explosive barks also have variations which likely have their special meanings.

Four of the pups were a mixture of black and brown, not true silvers like their father, and yet not of the red or cross-color phases. The fifth one was a typical red fox, with the distinction of a white brand on top of the left hind paw. He was tame and confiding, like the mother, while the other four with their black faces had a wild look in their eyes and remained scary and suspicious.

Split-ear yawning.

Family life centered on the five little foxes. At this time the mother nursed them at intervals. Between nursings she curled up somewhere near the den or wandered around in the vicinity, uncovering caches for a bite to eat and sniffing here and there. Not until the vixen's mate appeared and took over the watch did she lope away to stretch muscles and use her sharp senses for mousing and squirreling. The last nursing I noted was on June 24, and as the pups grew, they became more independent but remained close to the den area throughout the summer.

On July 6 the handsome male rested on a low gravel ridge; his moderately tired muscles were regaining full resilience while he baby sat. Split-ear had gone hunting, more for joy or habit than because of any need. The pups at this time were half grown and wandering about the den area as they wished, from one to another of the five sets of burrows some fifty or one hundred yards apart. Often they disappeared into the shallow, brushy ravine or wandered out on the dwarf-birch flat a short distance. But now they were all in the earth. Northern twilight, raining a little, and cold. That, briefly, was the setting for the "live show" I was watching — whose plot was not yet written.

The handsome male mostly rested on his stomach, muzzle on the ground between forepaws. Once he became aware of something approaching, for he stood up and trotted and watched to the east. It was one of the extra red foxes that was making a casual call and it soon left. The male was not interested and returned to his gravel ridge. He yawned, stood up, made a turn, and lay down again. A fox yawning is not exciting, but somehow it pleased me, for it was intimate and natural and the action accentuated the stillness.

A few nights earlier, while at the den, I had scanned some distant low hills through my telescope and noted at their base a group of caribou looking intently at something which proved to be a wolverine galloping up the slope. Now, while

I was looking over this same slope on the chance of seeing another wolverine, the male fox disappeared. One of the pups in a frolicsome mood ran to the gravel ridge where the male had rested, and then continued into the dwarf birch beyond. Apparently, the male, his nose and ears always alert, had again become aware of someone approaching; or perhaps he had left to hunt again.

Then I heard the hoarse bark of a fox two or three times out in the brush, and presently saw the black male poised and tense on the edge of a bank, head pushed downward toward a burrow entrance and a blackish animal. My first thought was that the black form was another black fox, but then I saw that it was a wolverine. Two or three times the wolverine jumped back into the brush, but each time returned at once to the burrow entrance into which it was poking its nose. The aroused male fox on the bank was exploding barks at this notorious miscreant only a few feet below him.

Split-ear had heard her mate barking and apparently had interpreted it to mean serious danger, for she had left her hunting to come bounding across the den area to join him. The wolverine continued pushing its head into the burrow, but either the burrow was too small to enter, or the wolverine did not wish to leave his rear unprotected from the two foxes barking and threatening close to his stubby tail. He kept turning on the frantic foxes, but they always jumped nimbly away. A few times, however, in escaping they almost collided because of their reckless excitement and close quarters.

After considerable skirmishing at this burrow, the wolverine led the way into a miniature ravine and up the other side. The two foxes kept barking and jumping close to his rear.

The two foxes kept barking and jumping close to the wolverine's rear as they followed him.

The pup that had gone into the brush near the start of the excitement now came out in the middle of the den area and stood there, scared and uncertain. Finally, he chose to retreat to a set of holes farthest from the barking. The wolverine moved in a semicircle and five or six times sat up to look around. He was a large, beautiful, well-furred animal with a conspicuous whitish collar across the throat. He seemed suspicious, as though he had my scent. Both foxes sky-hopped when they lost the wolverine's exact location after jumping away from him. The vixen was the more prone to lose the position of the invader, for she was less bold than her mate, who followed more closely. When the wolverine and male fox again entered the ravine, she lost them in the brush and came dashing to the near bank. I watched as she raced silently and intently along the rim. When she came opposite the two in the ravine, she barked and jumped about in a frenzy on the bank above them. Soon the wolverine returned to the burrow and again tried to enter, once making a few scrapes with a paw. But he must have decided that the project was not feasible, for soon he moved into the brush, escorted by the male. The fray had lasted about fifteen minutes.

Forty minutes later, Split-ear began looking for her mate, and in five minutes she was able to run joyously to him. He had, no doubt, followed the wolverine far across the tundra, as I once watched another fox do. Now he dropped a mouse he was carrying, and Split-ear galloped to the old den with it and called to the pups. The male spent some time sniffing at the burrow that the wolverine had threatened to enter. Soon he disappeared, perhaps to hunt. I then walked over to the burrow and measured the entrance; it was approximately six inches wide and eight inches high. Split-ear accompanied me and gave the place a nose inspection. It was nearing midnight and a light rain was still falling.

One morning on Sable Pass, when I was on my knees examining closely the vegetation to learn what plants a grazing grizzly bear had eaten, I heard a slight rustle behind me and, on turning my head, was a little startled to see a red fox five feet away. It was a friendly male setting out on a hunt. He had no special interest in me and passed by, stopping to sniff the delicate odor of a clump of chiming bells, as though enjoying the tundra flowers — a beautiful thought — but the next moment he shamefully substituted his own scent. Why do foxes and wolves, along with some other species, leave their scent on clumps of vegetation, a bone, or similar object! Seton, the artist-naturalist, suggested that a scent station serves as a newspaper — posting the news as to who has been that way — a register for signature.

The male continued on his way, with a carefree, exploring mind, ever alert, enjoying it all in the spirit of adventure. On all sides he was greeted by the alarm calls of the ground squirrels standing erect beside the safety of their burrows. Aware of the handicap of all this broadcasting, he took advantage of depressions and other cover as he traveled. He moved along, knowing that opportunities for pouncing on a squirrel could come at any time.

Once I saw a fox make a quick dash to one side for a squirrel; by coincidence he almost ran over a willow ptarmigan, which in sudden alarm took wing barely beyond his jaws. The fox was also startled and did not even try to reach for it. But he captured the squirrel and stood for a few moments panting, with jaws open. Not being hungry and not wanting to carry the squirrel with him, he pawed a hole in the sphagnum in which to cache his victim, changed his mind, dug another hole, inserted the squirrel, and covered it by pushing moss and debris over it with his nose. When I examined the cache the next morning, I learned that the fox had returned and eaten the anterior half of the squirrel, the preferred part, and recached what remained. Outside the cache he had carelessly left a hay mouse.

As he treads his way, the red fox often stops to listen to a mouse hidden in the thick grass cover. Sensitive ears and nose determine the mouse's exact location. He poses, watches, sways perceptibly with tense muscles, makes a ten-foot pounce, and the mouse is pinned beneath the grass.

The fox afield has many little adventures. When he passes into the nesting area of the long-tailed jaeger, he is mobbed and hurries away, but he jumps into the air at the birds, seeming to make a game of it. And likewise the short-billed gulls set up a bedlam when he comes near their nesting pond or river bar. A golden eagle may circle low, apparently a threat, for the fox stands with tail erect like a ramrod, on the watch until the eagle circles away. He knows the grizzly and often spends time watching him, especially if the bear is near the fox den and needs leading aside; and he may even play with a bear cub. A wolf, he may follow closely at times, and with security, for he can easily escape. He is acquainted with everybody on the tundra and has them all appraised.

Hunting may take a fox three or four miles from the den or, of course, a much shorter distance. Sometimes, he accumulates quite a mouthful of food items which, if he continues hunting, he deposits on the ground each time he hears or smells another potential victim, in order to be ready to pounce. One day I saw a fox that was carrying a load deposit it in the shade of a bank and lie panting. After a short rest, he gathered up his booty, a squirrel and some mice, and continued on his way to the den. There he would arrive ready for a prolonged rest after the joy of being afield. What a variety of tundra news is tucked away in his memory!

13. Happy Foraging

AT A PROPER DISTANCE from the hotel at Mount McKinley National Park is the dump. It is apparently rated by many furred and feathered folks as one of the best in interior Alaska, a reputation it has gained because of the quality and quantity of the garbage ingredients, and their regular daily delivery. In general appearance the dump is like many others. There is an assortment of wooden boxes, fresh and soggy paper cartons scattered down over the edge, old and new cans, discarded clothing, and ashes, which are used for surfacing as the dump builds out from the original bench on which it was started. To one side, permanently parked, is a collection of miscellaneous discarded road machinery and worn-out trucks. Out in front is a flat of scraggly spruce that breaks off into lower flats to the Nenana River and to miles of Alaska wilderness. A geologist would tell you that the dump is located on an ancient bank of the river.

I do not remember when I first became aware of the potentialities of the hotel garbage heap, but I do remember that for a time I did not know its location and was not particularly anxious to find out about it. I think it was the ravens that first attracted me to it. Anyway, it developed that I was making visits there at odd times and becoming intimate with its society. I later learned of other garbage dumps. There are many of them along the Alaska Railroad, probably one near each sectionman's station. I understand that an excellent one is at the railroad town of Healy, ten miles away, and a college boy who worked one winter at the coal mines at nearby Suntrana told me of a good one there. Refuse heaps are characteristically present at mining camps and at the scattered cabins and settlements throughout interior Alaska. They constitute a feature of some importance in the wildlife environment of the country.

Each garbage heap is the hub of a small bird and mammal community; an inadvertent feeding station. Following the lines of least resistance in gaining a living, many species of wildlife visit the dump. I have always looked a little askance at this sort of thing, this free-lunch counter, for it is somewhat like a dole. There have been cases where animals depended too much on handouts and suffered physically and, from the human standpoint, esthetically. One does not like to see a bear living entirely on handouts, for he is much more attractive and significant rambling on his own. Therefore I was pleased to learn that the animals at the refuse dump did not rely entirely on it for a living. For example, the ravens, although they fed there extensively, ranged far over the countryside and could be seen on the high, bare ridges of the mountains, awkwardly hopping about, feeding on the red cranberries. When I found a moose carcass six miles from the garbage, and another one ten miles away, I met the ravens at each one, but I knew they had been at the garbage because of the wadded paper and pieces of dishrag in their regurgitated pellets. The garbage did not cause them to become lazy, for they continued enjoying the active pleasures of the raven world. Likewise the magpies and the Canada jays did not entirely forget the ways of their ancestors

but continued to feed on the normal winter foods, largely
berries. For most of the species this artificial food serves as a
supplement to the regular diet. Only when the snow is especially deep, or when hunting is poor, might the garbage become a crucial factor for survival. It seems likely that some
wolves in recent years have depended rather heavily on refuse heaps. Being highly mobile, they can readily visit and
benefit from a number of them. But not always do they find
the fare of a high order. The last meals of two wolves shot by
an Indian up the Nenana River consisted of the leather tops,
including buckles, of a pair of shoe packs. The leather had
been chewed into strips and was in a compact ball in each
stomach.

oJ·M·

Even the wolf may follow the way of least resistance and visit the garbage dump

Usually we think of the garbage as mainly a winter feature, and it is. In summer, the raven and magpie and other winter residents disperse to raise their families and live in the wild. The ravens fly off to distant cliffs for their nesting and are scattered so widely they are not often seen. But the dump is not entirely neglected, even then. At this season it is largely taken over by the short-billed gulls, who arrive early in the spring when the snow is melting. Occasionally a grizzly, easily enticed by the smells and flavors, makes a visit, but he is seldom seen. A black bear, rather rare in this high country, was once seen stealthily moving through the willows, poking his head into view at different points along the margin, to spy and learn if the coast was clear. Later a black bear became a regular customer, but his droppings indicated that he subsisted largely on the bitter buffalo berries during the latter part of the summer. It is especially desirable that the bears remain in the hills: when they associate with humans they tend to lose their good behavior and respect, they become unafraid, and problems arise because of the bears' great strength. Incidents take place that are often unpleasant for all concerned. We prefer seeing a bear on a slope, digging his ground squirrels and grazing on grass and berries.

There is usually a band of gulls on hand, thirty or forty of them. They assemble from far-away nesting sites, for I have watched a single gull or a pair of them ten or fifteen miles from the dump, winging steadily eastward in a direct line toward it. Only occasionally would one turn aside briefly to inspect something more closely. At the dump a few large, ungainly, yellow-eyed herring gulls would stand among the smaller short-billed gulls, mostly tolerating them but occasionally administering a nip, sometimes sharp enough to loosen a feather or two.

One evening after a rain we watched a gull carry a piece of toast to a puddle and, without the least embarrassment, dunk it in the muddy water.

One evening after a rain we watched a gull carry a piece of toast to a puddle and, without the least embarrassment, dunk it in the muddy water. The gull swished the toast around in the water until soft and then swallowed it. As we watched, other gulls did the same thing, except that there was some variation in the manner of dunking. The toast was sometimes dropped in the puddle and allowed to sink to the bottom. That was not the best technique, because the water was so muddy that the gull could not see the toast and had to prod around to find it. By the time the morsel was found, it was so soft it disintegrated. These muddy puddles were also used for washing bills and as a source of drinking water.

By the beginning of winter the gulls have flown south. The most conspicuous resident now is the raven, a rugged, picturesque bird, a valued compatriot, worthy of special admiration and protection. Among those keeping company with the raven are his intelligent relatives the magpie and the Canada jay. One day a gyrfalcon appeared, causing the magpies hurriedly to seek refuge in the willows. Two ravens tormented him until he flew swiftly away. But the gyrfalcon was probably more interested in a warm meal of magpie than in the refuse.

When one approaches the dump at this season, the black ravens flap away with powerful wing beats. Some alight in scattered spruces to wait, others soar and play in the wind, and still others go out on the flat to investigate bones and other items, mostly carried away and left by foxes. The magpies, conspicuous in black and white, fly off to the willow thickets, trailing long, iridescent tails. They seem quiet and subdued, probably because they are occupied in examining all those shiny cans, and some not so shiny, and poking about among the paper cartons and wooden crates. The camp robbers, always tame and confiding, may not even fly away, and should there be a rare flock of snowbirds, they either take short flight to a nearby knoll or, as fancy strikes them, fly off into space. Always there are a few magpies out of sight among the cans, too engrossed to leave until one looks over the edge at them.

The behavior of the ravens at the dump suggests that they mate on a permanent basis. There is much pairing off, both in flight and when they are perched in the spruces. On one occasion in November three pairs were definitely petting; I assumed that they were pairs, for in each case one of the ravens mildly strutted and raised the long throat feathers in the manner of the male, and the other seemed coy and ladylike. One of the strutting birds faced his companion with open bill, then nestled his head under her bill and against her throat, in the spirit, as it seemed, of holding hands. Another pair was so perched that when one of them looked upward her bill was almost touching her friend's toes. She seemed to be wanting attention, for she held her head up for some moments in a stiff pose. Her friend complied with her wishes, apparently, for he proceeded to ruffle up the feathers on the top of her head, worked his bill down on both sides of her face, and then ruffled up her throat feathers. This continued for several minutes until the one getting her head rubbed made a move as though to peck her kind friend, and hopped off her perch into flight. In the case of the third pair, one of the birds fondled by her companion's bill also wearied of it after a time and administered a sharp little peck on his back. That little peck apparently meant "enough." She flew away to a patch of willows and came up with a piece of white garbage.

Ravens have a number of distinct calls, but never did I hear such a variety as on the day I circled below the dump to check on tracks. As usual, the ravens had flown away at my approach, and when I got out of the car and walked toward them they flew out of view — all except one. He was over in a green spruce, acting as though quite displeased with me. He made a series of sharp noises by snapping his big bill; he followed these with the common croaking sounds, then shifted to a rarer, low, buzzing sound and then to the typical, musical, bell-like, clunking notes. The calls shifted from one type to another; he repeated the repertoire over and over, not necessarily in the same sequence. He flew nearer to me, and the variety of calls became even more amazing. He croaked on a high pitch and after an interval abruptly switched to a deep bass. Then the articulation changed, and he cawed like a crow. He made at least a dozen distinct calls. The variety was obviously not due to a careless performance or to inadvertent deviations. It was raven talk. He was telling me in every way he knew that the arctic day was short, that he must complete his afternoon meal to fortify his system against the long, chill, arctic night, and would I depart so that he could safely return to that delicious morsel he had discovered just before I appeared. At intervals one of his companions sailed overhead to see if all was clear and, upon unexpectedly discovering me close below, beat the air loudly in a hasty turn to get away. On this occasion I returned to my car and backed off a little way.

Now somewhat normal activity resumed among the broken boxes, cans, and debris. A magpie, with legs braced, hammered at a frozen morsel with partly open bill, his entire body behind each blow. There was a shattering of the surface and he picked up and swallowed the chips. What remained of the chunk, he lifted from place to place, as though considering what to do with it, unable to make a decision. He finally left it and made a search for something better.

A raven struck with its huge bill at a frozen mixture containing coffee grounds and then walked in a flat-footed, stately manner to something else. A magpie, returning from a caching trip, seemed to misjudge his speed in landing on the edge of a protruding board and tipped far forward before gaining his balance. (They usually make this kind of landing.) A few camp robbers sailed in lightly, old hands at salvaging camp foods of all kinds. They swallowed all they could and flew off with mouths crammed. Probably, like the magpies, they regurgitated most of the swallowed food and cached it.

A raven was using a dead branch near the top of a spruce for a perch. When he first had alighted, he rubbed both sides of his bill on a twig and then on his perch beside his foot. There was preening needed on the shoulders, the breast, almost all over. Plenty of time, no hurry. Finished,

A magpie, returning from a caching trip, seemed to misjudge his speed in landing on the edge of a protruding board.

he nestled close to the branch and fluffed his feathers to keep his feet snugly warm. Four other ravens alighted on a dead spruce and were silhouetted against the snow-covered pyramid peaks on the distant sky line. Another raven backed off the end of his perch (the stub of a branch) and swung, hanging by his toes. From this position he examined the perch minutely. A raven carrying a large white morsel was chased by four others until he dropped the morsel.

Buck, the thirteen-year-old sled dog now pensioned off, and the last of the forty dogs the Park Service owned before the war, followed my tracks to the car and seeing me, slunk away in his wild fashion.

A red fox, following a winding, beaten trail through the old "burn," trotted with his nose close to the snow to catch every smell. He came upon a big beef joint, familiar because he had chewed on it last night, barely succeeded in getting his jaws around it, wondered what to do with it, then dropped it, only to pick it up again. He concluded that there was nothing to do about it and left it in the snow. It was too cleanly gnawed to bother with, even though its odors were still a delight. There must be something better farther on, so he continued along the trail and came upon three magpies fussing around a cloth sack containing food they were unable to get at. They quickly gave way, hopping a few feet to one side. The fox gave the sack some shakes and jerks. Now eight magpies gathered in close, not wanting to miss out on anything. A scrap of some kind came free and the fox picked it up and trotted off to one side with it. He poked it into the snow and, with long, forward-pushing strokes with his nose, covered it with snow. A magpie followed him to watch, and soon he would rob the cache. I was hoping to see the beautiful silver fox, a regular boarder, but no doubt he was detained and would arrive in the dark. Another red fox approached through the open poplars but noticed the car and curled up on the snow, perfectly content to snooze a little longer before beginning his night prowling.

A red squirrel thought he owned the dump and chased another squirrel off the premises to a clump of white spruces. This second one showed that he need not resort to the garbage for a lunch. He climbed a small spruce, cut four inches off the topmost twig and, deftly manipulating the piece of twig with his paws, neatly removed each tiny bud. The terminal twig, so lately the most important one on the spruce, he casually dropped and it lay on the snow. The squirrels were born and raised here, so they thought the world was mostly garbage.

About three hundred yards from the dump is an old one-room log cabin, once used for a post office. The sign is still nailed above the door. The shelves that served as receptacles for the mail are still in place on the back wall. All available space on them was filled with dozens of bread biscuits. Biscuits, too old to serve the hotel guests, had been thrown away, and red squirrels with sprightly energy salvaged what they could in the competition and hoarded their booty in the old post office, where long-awaited letters from home once reposed. A year later the biscuits were gone, and all the shelves were crammed with mushrooms. Had the biscuits, enough to last several years, been thrown back on the dump?

Among the mammals there are two outstanding winter guests. Occasionally I see the typical track pattern of the loping wolverine, villain of many a trapper's story and steeped in tradition, who rambles widely over the snow country alone. He is an irregular guest, and I am rather surprised. One would imagine that the garbage would have a strong appeal for him, who passes up no carrion, but I am pleased that he mostly stays by his old way of life.

The biggest thrill of all is to see the large, five-inch tracks made by the timber wolf, coming out of the wilderness to seek food where he can find it. There are several trails he uses in making his stealthy approach. My brother, Olaus, who has mushed dogs on many long Alaska trails and cannot forget the Alaska wilderness, recently wrote me, "Gosh, if I could only follow a wolf track for a day, even if the itinerary includes the dump!" I can readily appreciate his feelings. The wilderness spirit associated with the wolf is strong enough to dominate any scene, even the garbage heap!

14. The Alaska Haymouse

O NE DAY, when I was in the hills observing the Dall sheep, I had to walk across a long slope of high tundra. My eyes were fixed on a feeding band of rams that I had already located on a distant ridge. But the lichen-covered hummocks and the low brush that made rough walking also required some attention, at least to the extent of quick downward glances as I picked my way and tried to keep from swinging a boot into a hummock or stepping with a jar into a knee-deep depression. Along the way an occasional ptarmigan cackled off in a short, low flight from where I had disturbed it as it gorged on blueberries.

Over to one side, in a swale, I saw a hummock that looked a little different. I approached more closely and observed that it consisted of piled-up vegetation. Here was something I had not seen before, at least out in the open country. Picking at a few samples of the vegetation, I recognized a species of fireweed; probing down into the middle of the pile, I saw that it was all dry and so well cured that it still retained its green color and fragrance. Some dweller of the tundra had been making hay, not in the accustomed hay season of July but in the rainy days of autumn. The hay had been piled to overflowing among the branches of dwarf birch which, to a certain extent, served as a rack or crib to keep it off the

damp ground. A few feet from the stacks was the patch of
fireweed from which most of the hay had been secured. The
stalks had been uniformly cut about an inch from the ground,
leaving the effect of a miniature stubble field.

I wondered who owned the hay. It would not have been
unusual or puzzling had it been located among the rocks,
where the cony was living and regularly making hay. Either
the cony was departing from his traditional rocks and trans-
ferring his activities to the tundra, or else some undiscovered
haymaker was operating. The cony I regarded as too set in
his ways to depart from the safety of his beloved rocks; and
such vegetarians as the ground squirrel, marmot, porcupine,
and snowshoe hare could be ruled out as unlikely. In looking
for sign, I found a few mouse droppings in the stack. I had
never heard of a haymaking mouse, but by elimination and
from the sign it seemed likely that one of the six or seven
species living in the park had this extraordinary habit.

I noted some landmarks so that I could again find the
place and continued on my way to spend a day among the
mountain sheep. After supper I returned to the swale with
a number of mouse traps in my pack. I should have liked
to dispense with the trapping and discover the haymaker in
a way more in keeping with my feelings for the mice, but
they are secretive, seldom seen, and hard to identify in life.
So I set traps at the haystack and in adjoining trails.

The following morning I eagerly approached the swale. A
few empty traps, and then a mouse. Was this the haymaker?
Had the mouse of such superintelligence and forethought
been discovered? Perhaps, but there were many more traps
to examine, any one of which might yield another kind. If I
were to be fairly certain of the haymaker's identity, all the
mice taken would have to be of the same species. To my dis-

ORDER FORM

To: Supt. of Documents
Govt. Printing Office
Washington 25, D.C.

Please send me _____ copies of the *DIRECTORY OF POST OFFICES*, at $2.50 per copy.

Enclosed find $_____ (*check or money order*).

PLEASE FILL IN MAILING LABEL
BELOW

Name _____

Street address _____

City, zone, and State _____

U.S. GOVERNMENT PRINTING OFFICE
DIVISION OF PUBLIC DOCUMENTS
WASHINGTON 25, D.C.

OFFICIAL BUSINESS
_____
RETURN AFTER 5 DAYS

PENALTY FOR PRIVATE USE TO AVOID
PAYMENT OF POSTAGE, $300

Your name _____

Street address _____

City, zone, and State _____

GPO : 1962——O—610083 #133

ADVANCE ORDER FORM

1962 Edition

DIRECTORY
of
POST OFFICES

Issued annually, this Directory includes the following lists:

A list of postal delivery zone offices

A list of all post offices, branch post offices, and stations arranged alphabetically by States

---

An alphabetical list of all post offices, branch post offices, and named stations

---

A list of post offices by States and counties

A list of number of post offices, by classes, in each State and territory as of July 1, 1962

A list of Army posts, camps, and stations, and Air Force bases, fields, and installations

---

A list of post offices that have been discontinued or had their names changed during the past 2 years

Catalog No. P 1.10/4:962

$2.50 per copy

Convenient order form on back →

appointment, the very next mouse was different, and by the time I had collected all the traps three kinds had been taken. Further trapping at this spot would be of no avail. Clearly the habitat was suitable for too many species. A place must be found where only the haymaker lived.

I noted a few additional haystacks that fall, but in view of the time it would have taken from other work with which I was occupied, all were in areas too distant to trap. When I finally left the park, I was still wondering about that unknown maker of hay.

Five years later, in 1945, when I returned to McKinley Park to continue my studies of wolves and mountain sheep, I still remembered the haymaker and was on the lookout for its little piles of hay. That year the haymaker was apparently near a peak in its population cycle, for the haystacks were much more plentiful than five years earlier. They were present in a variety of places, from the stream bottoms to the ridge tops. Many were in dense spruce woods where ground plant growth was sparse except for the deep, spongy layer of mossy vegetation; others were on the open tundra; many were in the willows along the creeks; and they were common among the mountain avens and heather of the high ridges occupied by the mountain sheep. Strangely enough, the stacks were absent from heavy stands of grass where one would expect conditions for haymaking to be ideal. But perhaps there was not enough variety here, not enough herbs, or possibly the haymouse shied away from another species of mouse that was abundant in this habitat. With stacks in such a variety of places, it seemed likely that in at least one of them only the haymaker would be found.

As a beginning, I casually set traps near my cabin, and here again took three kinds of mice — the redback, the tun-

dra vole, and the Toklat vole. My search was narrowing
down. The mysterious mouse must be one of these. As time
permitted, I trapped a number of places. In those containing
a mixture of vegetation, including considerable grass, the
tundra vole was present with the Toklat vole and an occasional redback. But, eventually, as the trapping continued,
I found that the Toklat vole was always associated with the
hay caches and, in many places where the caches were plentiful, only this mouse was found. It soon was evident that the
Toklat vole, whose scientific name is *Microtus miurus oreas,*
was the thrifty northerner who shared with the cony the
distinction of making high-quality, fragrant, green hay. The
range of the species lies in Alaska and adjacent parts of
Canada.

At first glance this appears to be just another field mouse,
one of the numerous varieties of the common meadow vole,
or field mouse, that inhabits our fields and meadows and
miscellaneous grassy localities from the Arctic to the deserts,
from the Atlantic to the Pacific; a tribe of lowly rodents that
pursue their subterranean ways under the matted grass or in
underground galleries. The vole's fur has a yellowish-brown
cast. Its eyes are small, black, and beady, not large like those
of the more nocturnal deer mouse. Its disposition is unusually gentle. When I placed five or six strangers from
different colonies together, they were perfectly friendly.
Perhaps the community enterprises have brought about an
unusual degree of tolerance in these mice.

Although the haymaking mouse's identity was now established, my interest in him continued. I found and examined many hay caches. They vary in size from a handful
to a heap that would fill a bushel basket. Curing of the
vegetation takes place in the stack, the method generally

used by the cony, and the green cuttings are added slowly
enough so that they can cure before they are too deeply
covered. For most types of vegetation no great care is necessary for curing the usual small additions, but in the case of
succulent plants, such as the coltsfoot, the leaves readily turn
black if they are piled too deeply, especially in rainy weather.
The cuttings are brought to the caches in lengths up to
fifteen inches.

I found the following plants in the stacks in large quantities, sometimes a single one of the species making up an entire collection: fireweed (*Epilobium latifolium*), horsetail
(*Equisetum*), coltsfoot (*Petasites*), willow (woolly- and smooth-
leaved varieties; *Salix*), mountain avens (*Dryas*), lupine (*Lu-
pinus*), and sage (*Artemisia hookeriana*). Smaller amounts of
pyrola (*Pyrola*), arctous (*Arctous*), and alder (*Alnus*) were
sometimes present. Grass (*Calamagrostis canadensis*) was occasionally stored and made up the larger part of a few caches
noted. Not all these plant species were always available to any
one group of mice. In some of the woods the principle food
to be had was lupine; at other places coltsfoot, horsetail, and
willow were all abundant and much used, while on the ridges
the hay consisted mainly of mountain avens. On these ridges
the evergreen heather was eaten but, so far as I could determine, was not stored as hay. Since it was evergreen, curing it
was not necessary. In places the mice had climbed to the
upper branches of willows four or five feet tall to clip off fine
twigs for storing. The wide latitude in this vole's tastes, together with its haymaking, has been conducive to its presence
in many kinds of habitats.

The most remarkable feature of the harvesting was the
care taken to keep the hay dry, especially at the base of the
stack. Among willows and other shrubs the hay was often

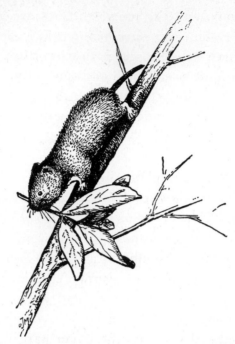

The haymouse also harvests the leafy twigs of willows.

placed in the basket formed by the limbs as they spread upward from the ground. Sometimes it was placed on horizontal
limbs, with occasionally a second limb lending protection
from above. On the open ridges, small caches were placed in
protected rock niches. In the woods, where hay was often
piled against the trunk of a tree, it was always placed on an
exposed root buttress rather than on the ground. The space
between a log and the ground was often used for storage; in
one place a space under a log was so used for a distance of
six feet. One cache was placed on a tangle of dead branches
fifteen inches up the trunk of a tree. Two of the branches
leading to the ground were apparently used for a bridge over
which to carry vegetation. The amount of hay stored by different colonies varied considerably, some seeming to depend

more than others on this type of food. Even when a large amount of hay is stored, apparently it is all used up during the winter if the mice have survived. Two colonies that I knew about were well supplied with hay in the fall but had consumed it by spring. Where each stack had rested, only an accumulation of mouse droppings remained. The snowshoe hare had raided several of the stacks, as I could see by the presence of its droppings at the storage sites. When the hare is abundant it might be a problem to some of the mice, but its depredations would be curtailed by a few falls of snow.

In places the mice had climbed to the upper branches of willows four or five feet tall to clip off fine twigs for storing.

One drizzly day, when the fog hung so low on the mountains that to pursue the regular field work with which I was engaged would have been unprofitable, I wandered up the bars along Sanctuary River. In the fine glacial sand I noted tracks of various kinds, including those of the grizzly, cari-

bou, wolf, and fox. Thinking of the Toklat voles, I wandered over to an older river bar on which vegetation had
long ago taken hold. Here a colony of voles was flourishing,
and I noted that there had been much fresh digging in the
fine sandy soil. While I was idly examining the structure of a
tunnel and following it to see if it would lead to a nest, my
fingers came to a place where the tunnel abruptly widened.
Lifting up the sod, I exposed an underground treasure house
packed to the ceiling with choice, fleshy roots. These mice, in
their wisdom, knew that hay keeps best above ground, roots
in the ground. Now I understood why I had seen so much
evidence of fresh digging. At the mouth of some burrows I
had noted as much as two or three quarts of excavated soil.
And the why of all the little prospect holes, two or three
inches deep, also became evident. Each hole represented a
harvested root. With my tape I measured the storehouse. It
was roughly sixteen inches long and five inches deep and
wide. The bountiful food supply gathered here consisted of
425 pieces of the thick, dandelionlike root of the legume
Hedysarum, 340 yellow roots of *Pedicularis* (which resemble gnarled carrots), 380 bulbous, J-shaped roots of the knot-
weed, *Polygonum*. This cache was six feet from the grass nest,
among the roots of a clump of willows. No doubt there were
other root cellars, and above ground there were twelve stacks
of hay in an area fifty feet across, so these mice had provided
well for the winter. By live-trapping I later learned that
there were five or six mice in the area, apparently a family,
for some of them were no more than half grown.

In another area, a chamber eighteen inches long, ten
inches wide, and five inches deep was filled with forty-five
horsetail tubers, 1,021 underground buds of coltsfoot, 155
pieces of coltsfoot roots, and 2,808 grass buds or shoots.

Three feet from this cache was another chamber of the same size, freshly dug, but as yet containing no stores. No doubt both caches belonged to the same mice. Several of the moss-covered hummocks in a wettish area were more or less hollowed out for caches. The main chamber of one such cache contained coltsfoot roots, while a compartment off to one side was filled with horsetail tubers. In a hummock where horsetail abounded, a long, crescent-shaped cavity, freshly dug, contained only a handful of horsetail tubers. In a wet depression, where the soil was saturated with water, the mice had dug around and removed the large roots of a robust sedge. This must have been quite a difficult and muddy job, but perhaps the big roots were worth it. At this time, September 20, the root storing seemed to be at its height and haymaking had slackened off.

It is evident that the root business takes much more of the mouse's time than the haying. First the storage chambers have to be excavated, and when they are finished, hundreds of roots are needed to fill them, and each root entails a certain amount of digging. The mouse's method of removing it depends upon the growth habit of the root. Vertical ones that grow like a carrot were dug loose and removed from the surface, while the long, horizontal ones and those that formed a complicated, underground network were secured by tunneling. The roots of the common fireweed, sandalwood (*Comandra*), horsetail, and coltsfoot were gathered in tunnels, and shallow, perpendicular roots such as those of *Pedicularis*, *Polygonum*, and sedge required no tunneling.

But this was not all that this vole of the northern mountains had contrived for its safety and welfare. I was thrilled at finding yet another sign of ingenuity, perhaps the greatest of all. Other animals make hay, others store roots. But our Tok-

lat vole also has what appears to be a unique burrow construction hitherto unknown. Many of the burrows are a
series of cavities joined by narrow apertures barely large
enough for the passage of the mouse. What is the purpose of
this burrow architecture? To describe the burrow, let us imagine how a mouse chased by a weasel would benefit. The
following incident is based on an excavated group of tunnels
leading to a nest, and is conjured from probabilities. . . .

With long bounds, a weasel of the year came leaping over the mossy floor in a
fringe of spruce woods.

With long bounds, a weasel of the year came leaping over
the mossy floor in a fringe of spruce woods. He was not a
great deal bigger than a mouse, but was sinewy and powerful
and the enemy of all mice. He stopped, sat on his haunches,
the lithe body erect but graceful, the black nose pointing forward from the top. The nose imperceptibly sniffed and
caught the aroma of Toklat vole. He dropped on all fours
and his slim body arched along in a series of jumps. Soon he

was following a path over which only seconds before a vole's feet had left rich scent. A few jumps, and a surprised vole was almost captured before it could scurry away. It was but a few feet from a hole leading into the sandy earth and into this it darted, a brown streak. The weasel followed, dashed into the hole and, being inexperienced in Toklat vole hunting, thought he had a cinch. He had captured several mice under just such circumstances — however, they were other species. The Toklat vole knows how to make hay, but also he has invented an anti-weasel device.

In the underground blackness the weasel found the way blocked; no, not quite, for the tunnel had been beveled to a small opening into which his nose would only go as far as his eyes. He dug rapidly with sharp claws and broke through to a chamber about three inches in diameter. A second frustration met him here, for on the far side of this chamber was another tiny opening. This he likewise tore open sufficiently to permit his slim body to squeeze through into still another chamber. And here again, a third impediment! After pawing through eight alternating chambers and constrictions, he reached a room containing a shredded grass nest, but of course no mouse. The vole scent was strong and caused the weasel to sniff about in the inky blackness. A two-inch passageway led to another chamber, but it was the toilet, with a crust of nitrate an inch and a half thick on the floor. He lingered but a moment and was soon poking his nose into two other tunnels that left the nest chamber. Following the one the mouse had taken in escaping upward, the young weasel again met one constriction after another. He had never experienced such tunnels. If his head were but a size smaller, he could slip through after the mouse without wasting time enlarging the openings. But the constrictions measured only three eighths to one half inch in diameter, a size too small for him. Like thousands of other weasels, he had his first lesson in the tunnels of the Toklat vole. He learned that this vole must be surprised away from its underground home, out in the labyrinth of exposed or moss-covered runways. . . .

The constrictions in these extraordinary burrows were most numerous in tunnels leading to nest chambers — logically, since their primary purpose is protection. Some short passageways leading to root caches were simple, but one that was five feet long had two narrow necks. A number of constrictions in such a burrow would make it difficult to bring large roots to the cache. How the specialized habit of building these constrictions developed, what the mice have in mind when they build them, is difficult to say, but whatever was their original purpose, they would seem to function efficiently as antiweasel devices.

As I became more familiar with the voles, I often visited some of the colonies I knew about. A short wait, and I would begin to hear sharp rustlings in the vegetation and catch glimpses of the mice scurrying from cover to cover. To begin with, I felt fortunate in hearing their birdlike twittering and in getting any kind of look at a mouse, but as I continued watching I could see them doing things other than just scurrying. By mid-August, when here and there a spray of willow was already touched yellow, the mice were fully aware of the passing of summer and were busy with all their autumn activities. One mouse might be seen at the entrance of a tunnel, furiously pawing a spray of dirt between its hind legs; another might be emerging from a tunnel carrying one or two pieces of root and entering another tunnel leading to the root cache; and a third mouse might be carrying a leaf to the hay cache, or visiting the cache for a premature sampling. I doubt that there was any division of labor involved. Rather, I should guess that the individual mice, with democratic freedom, shifted from one job to another as the spirit moved.

A colony of mice near our home did an unusual amount of digging. This, I think, was due to the type of roots they were seeking. The roots available were the long, horizontal ones of the fireweed and Comandra. To get at them, the mice had to tunnel, and that required much excavating.

Often I saw the mice carrying the large coltsfoot leaves, which have a broad surface like a rhubarb leaf. A leaf eight or ten inches in diameter would sometimes catch along the sides of the trail, and when this happened, the mouse would get out in front to tug and pull like an ant. When free, the leaf was quickly carried forward again, the leaf much more visible than the mouse. One of the mice seemed to have used its ingenuity in this matter, for he had folded a leaf a few times so that it looked and carried like a well-wrapped package.

At one colony, at the beginning of the haying season, I saw mice that seemed to be examining storage sites. A half-grown youngster examined a favorable location in the heart of a willow clump, and at least three times I saw a mouse on an elevated horizontal snag that would "fill the bill." One of the mice on the snag, standing upright on his hind legs, with one paw free and the other resting on a stick, seemed to be sizing up the situation with considerable circumspection. The slight sag in his paunch made him look like an alderman.

The mouse colonies were by no means permanent. One that was very active storing hay in the fall of 1946 and had consumed the hay during the following winter had disappeared by the fall of 1947. A colony that I watched in August for a few nights was suddenly wiped out or quit the area. It had cut many coltsfoot leaves and was an unusually active group of mice. I later wondered if this activity and excessive twittering I had heard indicated a nervousness or restlessness that was connected in some way with the disappearance of the colony.

The Toklat vole has many enemies in the fox, wolverine, wolf, and birds of prey, all of which no doubt appreciate its contribution to their bill of fare. Even the grizzly counts the Toklat vole as one of his benefactors, in a small way, of course. Primarily a vegetarian, the grizzly feeds mainly on grass during June and much of July. Then, when blueberries, crowberries, and buffalo berries ripen in late July, he concentrates on them. But in the spring and fall in Mount McKinley National Park he must turn to roots, and with his powerful arms and long claws he pulls loose chunks of sod and, delicately pawing away the dirt, feeds on the roots exposed underneath. Imagine a grizzly's pleasure then as he comes ponderously on his way, swings his huge head over to a hummock and there, investigating a mouse hole with his nose, catches the sweet odor of choice roots. Having scented the roots, he straddles the hummock with forelegs, grasps it with both claw-armed paws, and, with a backward tug or two, easily pulls off the top, exposing two or three quarts of choice roots, some of the kind he himself digs and others that he likes but which are too small to pay for his digging them. Sometimes a mouse in a nest, rather than a cache, will be ruthlessly exposed.

I know that the Toklat vole, along with other mice, has a practical value; he aerates and enriches the soil, and he is part of the important link that takes the product of the sun's energy given to plants and transfers it to food for an interest-

Where the Toklat vole keeps company with the handsome Spruce Grouse.

ing part of our fauna, that is the flesh eaters, such as the marten, wolverine, lynx, coyote, red fox, hawk, and owl, and a host of other remarkable animals. I know that the mouse helps make the world in this way more varied and interesting, and I appreciate him for this. But I also value him for himself, for his intrinsic interest. And I enjoy letting my thoughts go back to the many attractive spots where the Toklat vole lives. I like best to recall his cozy, elfish runways in the moss of the spruce woods, where he keeps company with the handsome spruce grouse. Here I think of him in the fall of the year, busy with his mousy affairs, making hay, digging and storing delicious roots, some of this kind, some of that kind, a good variety. I can see him pushing his little nose into a hay pile, pulling out a stem, and rakishly feeding himself with one paw. A companion joins him and they both utter soft, friendly, birdlike twitterings, forgetting for the moment the persistent weasel, the sharp-nosed fox, and the powerful paw of the grizzly.

15. Gulls and Mice

Eᴀʀʟʏ Jᴜɴᴇ, and all about me on the south-facing, open slope above Sable Pass was nuptial music. Pipits, finding the hummocks too low for the full expression of their elation, flew skyward and spiraled down on set wings, singing. The Lapland longspurs, more striking in their jet-black and chestnut colors, also took to the air and poured forth their song, the chiming notes seeming to fall from the sky like rain drops. The horned-lark notes from up the slope were subdued and sweet. Rock ptarmigan males, in white winter feathers still, were perched conspicuously on hummocks. Their guttural voices, pitched low like the voice of a bull frog or the deep-bass "b-a-a" of the Alaska sheep, seemed hardly appropriate for birds. Up the slope I heard the "throidee" call of a surfbird. A pair of snowbirds, flashing black and white, fed restlessly with many short flights. Rosy finches, very tame, foraged energetically. Ordinarily these last three species would be near the ridge tops on their usual nesting grounds, but spring was at least three weeks late, and they were adjusting to it by remaining on lower slopes. The immobile cranberries, I noted later in the season, were unable to adjust themselves to this late spring, for many were in flower in late July when, to be successful, they should have been in fruit.

Over on the snow slopes a dozen caribou passed by, moving forward on their annual migration, expecting to find better grazing farther along. But the expected grazing was not there, and the vanguard had to move to lower ground along the rivers to feed. Habits and habitat which usually were synchronized were not quite meshing this year because of the very late spring.

Below me, at a small tundra pond, was a group of short-billed gulls. Some stood idly on the shore ice, a few floated in the open water, and some were huddled together in a little group on a knoll beside the pond, apparently well fed and therefore contented. For a few days previously I had seen the gulls, and I wondered why they had assembled there.

The almost inactive gulls seemed much less interesting than the distractions around me. There seemed little hope of anything happening as I watched them in their repose. On the edge of the shore ice I did note, after a time, a gull by itself, picking at a mouse. Here was an indication that gulls might be feeding on the numerous mice which other mouse eaters, such as the foxes, short-eared owls, and hawk owls, were enjoying also. The brown lemming and the haymouse, two of the half dozen species in the park, had prospered and had multiplied rapidly, like some human populations. Favorable hummocks were riddled with holes, and from where I sat I could see the mice scurry across short pieces of exposed runways. There was much fresh excavation — the big population was preparing to become bigger. There was no smog problem, but congestion would no doubt create a big die-off before many months.

My attention shifted to a lone gull that had walked seven or eight yards up in the snow and stood facing away from the pond. He (assuming that it was the male of a pair) was

She grasped the mouse and carried it to the water for soaking.

joined by a gull which I took to be his mate because of the behavior that followed, although I could be wrong in this. She assumed a stance facing him and pushed her bill close to his. This seemed at the time to be part of a billing performance. To avoid her bill, the male swung his head from side to side, and I wondered if this was a nuptial ritual. But it proved not to be exactly that. It was only the act of an insistent wife demanding the Saturday pay check from her husband before it is spent. It was no doubt an old custom, for the male seemed to know that there was no recourse and he made no real objection to complying. Apparently as soon as he could, he "ponied up," regurgitating a complete mouse he had been carrying in his gullet. He continued standing stoically as before, while she grasped the mouse and carried it to the water. There she sloshed it about, dunked it, and picked at it. One might assume that she was fastidious about eating a vomited mouse. She swallowed the wet, well-washed mouse whole, and I wondered why she had washed it so thoroughly.

There was a lull in feeding activities after the disappearance of the disgorged mouse. A long-tailed jaeger lit in the

She assumed a stance facing him and pushed her bill close to his.

water and proceeded to bathe exhaustively and with vigorous flapping of wings, dipping and shaking of feathers. Two gulls swam toward the jaeger in an unfriendly attitude, judging from other observations. Perhaps the power of suggestion affected the gulls, for instead of bothering the jaeger they began to bathe, too. The jaeger, for his part, was too busy with his bath to worry about the gulls.

Up in the snow was another gull that had "gone to the rail." Soon he was joined by his mate and, as in the performance of the first pair, heads weaved back and forth as the male seemingly tried to avoid the female's bill. She did some screaming, perhaps because of impatience. I knew what to expect this time, but it didn't work out quite the same way as before. He tried to disgorge a mouse, but couldn't. The mouse came part way up, and I could see its rear end in the depths of the gull's open mouth. But it would come no farther. It went down again. There apparently was not enough lubrication. The dry mouse was sticking in the throat. And that seemed to be what the gull concluded, for he flew to the pond and swallowed four or five gulps of water. Then he suc-

ceeded in ejecting the mouse. Not forgetting his mate, he
flew to her side and dropped the mouse before her and she
immediately swallowed it.

I was beginning to get the significance of what was happening. A dry mouse could be swallowed, but it tended to
stick in the throat. The female that seemed to wash her
mouse before swallowing it was not concerned about clean-
ing it but about wetting it so it would swallow easily and
continue readily its passage through the digestive system.

A third begging scene took place, but there were no results. Apparently the male was not prepared to serve a mouse
just then.

Other incidents followed which seemed further to indicate that mice were soaked in water for lubrication. A gull
walked out on the snow, disgorged a mouse, carried it to the
water, and after wetting it thoroughly, reswallowed it. Three
other gulls went through the same procedure. Mice swallowed in the field were disgorged for a soaking and reswallow-
ing. The soaking seemed desirable to lubricate the mouse for
passage beyond the upper gullet.

In these instances the gulls had not disgorged mice until
at least some interval after their arrival at the pond. Now I
saw a returning gull disgorge a mouse upon alighting and,
after wetting the mouse, reswallow it.

Still another variation. On three occasions gulls came to
the pond, each carrying a mouse in its bill. Each gull proceeded to wet the mouse thoroughly and then swallow it.

Many gulls returned to the pond showing no evidence of
having a mouse on their persons. But they probably were
carrying concealed mice and would later go through the ejec-
tion and wetting business.

All the gulls upon arrival at the pond seemed thirsty. They

lit in the water and immediately gulped down several mouth-
fuls. They may not have been thirsty in the usual sense but
were drinking to soak up a mouse in the throat. After the
drink, many of them bathed.

Toward noon, activities slowed down and more gulls just
stood around roosting. A few were up on the broad snow
fields, making short runs here and there and catching insects.
Obviously the gulls were benefiting from the high mouse
population in the land. They had found the pond conveni-
ently placed and were using it for their dunking require-
ments.

The rendezvous became less popular in a few days as more
snow disappeared and water was more widely available. I
searched the shore line of the pond and the roosting site on
the knoll and found many pellets ejected by the gulls; they
all contained remains of undigested fur and bones of mice.

The mouse dunking seemed to be a well-established habit
among the gulls I was watching and might be quite common
among gulls of this and other species under certain circumstances. There is a record of a herring gull wetting a rat before swallowing it, and there may be other similar records in
the literature. But perhaps large gulls, with large gullets,
manage mice readily without wetting them.

I was especially interested in the mouse dunking because
of its relation to my earlier observations. On at least three occasions near the hotel dump I had watched the short-billed
gulls carry hard bread and toast to water for soaking before
eating. This, at the time, seemed a rather quick solution to a
problem and suggested a high intelligence. After watching
the gulls wet the mice, however, it was apparent that the
bread dunking was more spontaneous, much nearer a natural,
age-old habit of the gulls, than I had suspected.

But my brother, Olaus, tells me that he has observed a grackle wetting a hard piece of bread before eating. Does the grackle, too, have a natural habit that is similar to the bread-dunking action? Or perhaps many animals have an awareness of the uses and qualities of the water element in their environment. They know that it serves for drinking, bathing, and washing food. The raccoon, extricating food items from mud, acquired the habit of using water for removing mud, and the practice became so strong that he might wash food whether it needs it or not. I have watched beavers dig roots out of soft mud and carry them to water for washing. This use of water was not as automatic as by the raccoon. Some animals know that water is good for softening food, a quality in water that they have learned to use. Wetting mouse fur is perhaps equivalent to softening food, or is it in another category — lubrication?

In the morning, when I arrived at the pass and heard the singing of the tundra bird life, I little suspected that the gulls, sitting so quietly, could vie in interest with all the other lively activities. Yet they had revealed what is no doubt an ancient habit adjusted to special needs: for centuries they have been wetting their mice.

16. In Search of Wolves

WHEN ONLY INDIANS and Eskimo knew the north, the wolf, caribou, Dall sheep, and moose existed together in interior Alaska year after year, century after century, each following its own way of life and each adjusted to the presence of the rest of the fauna and flora. Today these species continue to exist together in the north country. The presence of the wolf adds immeasurable richness and a wilderness spirit to the landscape. One need not see a wolf to benefit from his presence; it is enough to know that there is the possibility of discovering one on some distant ridge. It is enough to know that the wolf still makes his home in this beautiful wilderness region to which he contributes vividness, color, and adventure. Many Sourdoughs officially cherish him as an emblem of unspoiled country.

The wolf experience I like best to recall happened one day in January some years ago. In the morning it was a crisp thirty-five degrees below zero. At a cabin on the lower Toklat River in Mount McKinley National Park, my companion and I started out at daybreak, he on snowshoes, and I on skis, each of us carrying a pack containing bedroll and food, enroute to Wonder Lake along the north boundary of McKinley Park. We were making a two-hundred-mile winter trip to carry out general wildlife observations. Our way led twenty miles up the

Clearwater River. Heavy frost covered the spruce trees. At intervals we encountered overflow water on top of the ice, which necessitated detours to avoid getting wet. Overflow is prevalent in extremely cold weather, for as the thickness of the ice increases, no room is left underneath for the normal flow of water and much of it is forced through cracks in the ice to the surface where it spreads underneath the snow. One tributary of Clearwater often builds up ice in this manner to a thickness of twenty or thirty feet. Enroute we noted tracks of many kinds — fox, wolverine, wolf, caribou, moose, squirrel, and weasel — and gained general impressions on wildlife presence and abundance. An interesting day. Toward evening it became stormy and we faced into a bitter wind blowing the first snowflakes. It became dusk, and by the time we left the river and turned in on a trail it was dark and stormy. In a few minutes we would have a fire going, a warm cabin, hot coffee, and a big meal. Now we savored the storm, for it made the cabin just ahead seem especially cozy. Then we stopped, transfixed, for out of the storm came music, the long-drawn, mournful call of a wolf. It started low, moved slowly up the scale with increased volume — at the high point a slight break in the voice, then a deepening of the tone as it became a little more throaty and gradually descended the scale and the soft voice trailed off to blend with the storm. We waited to hear again the voice of wilderness in the storm. But the performer, with artistic restraint, was silent.

I was a little uncertain about the identification of the first wolf I saw in Mount McKinley National Park. I was not sure whether it was a wolf or a coyote. The animal was some distance away, on a ridge parallel to the one from which I was watching, so that size was not a good criterion for identification and the color was about the same as that of a coyote. But I

noted that the legs appeared exceptionally long and prominent, and that there was something about the hind quarters
which was peculiar, giving the animal just a suggestion of being crouched. On the slope it appeared more clumsy in its
actions than a coyote. Its ears were more prominent than I had
expected in a wolf. These characteristics I later found to be
typical. The behavior of this wolf was of special interest, so I
checked my identification by an examination of the tracks,
which were approximately five inches long. After becoming
familiar with the wolves, I generally found no difficulty in
making identifications, but still on a few occasions I could have
mistaken a distant brownish wolf for a coyote.

Some northern sled dogs resemble wolves so closely that it
would be hard to identify them correctly if they were running
wild. A few years ago I drove a sled dog which, even on a
leash, could be mistaken for a wolf. This animal was supposed
to be a quarter-breed wolf, however. It was lankier and had
longer legs than the average sled dog. His chest was narrower,
the front legs much closer than those of the more broad-chested
sled dog.

Wolves usually are classified as white, black, and gray, but
among these types there is infinite variation. Wolves referred
to as gray are sometimes of a brownish color similar to that of
a coyote. One type of animal is whitish except for a black
mantle over the back and neck. In some, the fur on the neck
is strikingly different from the rest of the coat. Some gray
wolves have a beautiful silver mane, the tail tipped with black.
Black wolves often have a sprinkling of rusty or yellowish
guard hairs which create a grizzled effect, and many are char-
acterized by a vertical light line just back of the shoulder.

I judge that the adult, male Alaska wolf in good flesh generally weighs approximately one hundred pounds. Estimates

have been made as high as one hundred and fifty pounds and more. There is much variation in size, and exceptionally large animals can be expected. Judging from the wolves that I saw, the females are smaller than the males.

The tracks of a wolf can easily be distinguished from those of a coyote by their large size and the long stride, but probably they would often be indistinguishable from those of a large northern sled dog.

The front foot is larger than the hind foot, being definitely broader and in most cases slightly longer. The width of the track, of course, varies according to the speed at which the wolf is traveling. When the animal is running fast or galloping, the foot spreads considerably.

When I measured the pace of wolves (the foreward distance between the tracks of the two hind feet or the two front feet; a stride would be twice the pace) I found that in walking and trotting it varied from twenty-five to thirty-eight inches. In seven inches of snow the pace measured between twenty-seven and thirty inches. In six inches of snow, the pace of a track probably made by a trotting animal averaged twenty-nine inches.

In traveling through snow, a band of wolves will often go single file, stepping in one set of tracks. The hind foot falls in the tracks of the front foot. At other times the hind foot may or may not fall in the track of the front foot. To avoid deep, soft snow the wolves often follow a hard-packed drift or the edge of a road, where the snow is more shallow.

I frequently saw flecks of blood in the trails in winter, indicating that the wolves were subject to sore feet. In the case of the sled dog traveling in snow, especially in crusted snow, the hide on the toes is often worn off, sometimes causing the dog to limp considerably. If the crust is severe, it may become necessary to protect the feet with moccasins. The feet of the wolves are probably affected by the snow in the same manner as are those of the sled dog, but possibly to a lesser degree. In summer a wolf would occasionally develop a limp and later recover.

The history of the wolves in Mount McKinley National Park during the last forty years is, in a general way, well known. There is also considerable information on the prevalence of the wolf in interior Alaska, although much of this information is conflicting. There is often a tendency to report a great increase of animals if any at all are noted, so that, during a period when wolves were much less numerous than at present, there were many reports of their abundance. Sometimes the caribou, the principal wolf food, shifts its range and brings wolves along with it into new territory, where they are noticed and commented upon. I suppose the history of the wolves varies a little in different parts of interior Alaska but that the general pattern is similar throughout. Since there are 586,400 square miles in Alaska, of which 3,030 square miles are in Mount McKinley National Park near the center of the Territory, it can perhaps be assumed that in its broad aspects the status of the wolves in the park has corresponded rather closely with their status in interior Alaska as a whole.

Some time after 1908 the wolf population in the Mount McKinley region and perhaps also in other parts of the interior of Alaska was considerably reduced, probably as a result of natural causes. An old-timer who had hunted sheep in the McKinley region in 1916 and 1917 told me that he saw no wolves or wolf tracks there at that time. In 1920 and 1921 my brother, Olaus, visited a number of localities in interior Alaska in his travels by dog team and found wolves absent or scarce in most localities. In a trip through Rainy Pass and into the Kuskokwim country in the spring of 1922 he saw no tracks. Mr. Joe Blanchell, at Farewell Mountain on the north side of Rainy Pass, said that wolves were formerly common in that locality but had now disappeared. Because there were caribou all through this area, ample wolf food was present.

Olaus made a trip through the caribou country between Chatanika and Circle in the spring of 1921, and into the Chatanika region again in the fall. He found no sign of wolf in the spring, in the fall he saw only one wolf, and he heard wolves howling only once. Part of this region was used the year round by caribou. A year or two before, wolves had been reported more plentiful.

In 1927 it was reported that the wolves were becoming more numerous. A band of eleven was noted by one of the rangers. From 1928 to 1941 wolves were reported each year as plentiful in the park. From the records available, it appears there were no large fluctuations in the population during that period. Since 1941, especially after 1950, wolves decreased in the park.

The most probable cause of drastic decimation of the wolves in early periods is disease. Mange, distemper, and rabies are some of the diseases which may affect them. Alexander Henry in his journal refers to scab in wolves. On March 5, 1801, at Pembina, North Dakota, he writes: "A large wolf came into my tent three times, and always escaped a shot. Next day, while hunting, I found him dead about a mile from the fort; he was very lean and covered with scabs."

R. M. Anderson writes in 1938 as follows concerning mange in coyotes: "One young male coyote shot by Warden J. E. Stanton when the writer was with him in Cascade Valley early in September was very mangy, being so nearly devoid of hair from nose to tip of tail that the scabby and vesicular skin was plainly visible on every part of the body. Most of the half dozen coyotes seen in this area appeared to be afflicted with mange, and several wardens stated that many of the mangy coyotes lived through the winter, but that the worst cases usually died in the spring. This disease, and perhaps other causes, seem to keep the numbers down, and the reports of the superintendent of the park show that coyotes have decreased in numbers in recent years."

Seton, in 1929, says that rabies seems to break out among wolves at times. Alexander Henry (1897) in his journal relates the killing of a wolf at camp which was thought to have rabies. Seton gives several instances in which wolves seemed to have had rabies. This disease probably could cause a drastic reduction in a population.

A disease like distemper could no doubt spread rapidly in a large wolf population, especially since the animals travel in packs. Distemper has been known to wipe out entire dog teams, and it might affect wolves even more severely. Although the young animals are most susceptible to distemper, older wolves that have not been in contact with the disease might be more vulnerable than old dogs, which generally are considered immune. In 1924, Olaus lost a dog team after he had traveled from Nenana to Hooper Bay, Alaska. Another team made up of older dogs was not affected. In that year a large number of dogs in interior Alaska are reported to have died from the disease; apparently it was present in epizootic proportions. During April 1934 an outbreak of distemper is said to have prevailed in various sections along the Yukon and Tanana rivers. The reduction in wolf numbers some years ago may have been due to the large number of dogs that were brought into the wolf territory for transportation purposes, becoming an agent for the spread of distemper or some other disease. The decrease in wolves since 1941 has probably been brought about largely by poisoning and by hunting from airplanes beyond the park borders.

In 1939, the first year in the field, I made no special effort to hunt wolf dens, because so many other phases of the study required attention and because the finding of a den can be extremely difficult. But in 1940 I made some effort to find one or more dens.

In den hunting, one must turn detective and use every clue available. Unfortunately some of the clues may only serve to give one a false notion of the location. Seeing wolves, if one is fortunate enough to see them, may be suggestive. In the morning, wolf travel is likely to be in the direction of the den, and in the evening away from it. But if the wolf happens to be twenty miles from home when seen, the observation is of little use and may be misleading. Much wolf travel is miscellaneous and irregular, and one can easily get erroneous notions and become more and more puzzled and confused. Seeing a wolf resting somewhere is especially suggestive, for it may be lying at the den. Tracks are always helpful.

I was much interested in a bit of woodcraft that an Indian from the upper Tanana River used in finding dens. He said that at first when he hunted for wolf dens he did not know where to hunt, but in time he learned more about animals and now knew where to look. Then he illustrated with an incident to show how he went about it. Once, in winter, he saw spruces barked by porcupines. The next spring, when he and his brother were searching for wolf dens, he returned and, pointing to the porcupine sign on the trees, said to his brother, "Probably a den over there." They went to the porcupine barking he had seen in winter, and there was a wolf den as he had suspected. His reasoning: he knew that the porcupine that barked the trees had to have a shelter and thought that maybe the shelter was a wolf den.

This bit of ecology fitted into my own experience, for I found a wolf den that a porcupine had used in winter and many willows nearby that the porcupine had barked. The method is interesting but not sure-fire, because most porcupines do not live in wolf dens. But this lore can on occasion be the clue.

After a fall of snow about the middle of May, I saw wolf tracks on the broad gravel bar of East Fork River directly in front of the little log cabin in which I was camped. The tracks led both up and down the river. Since there was no game upstream at the time to attract the wolves, it appeared that some other interest, which I hoped was a den, accounted for the movement that way. I followed the tracks for a mile or more to a point where they climbed a bluff bordering the river bar, and there I surprised myself and a black wolf, a male. He ran off about a quarter of a mile into a ravine and howled and barked at intervals. Then, following tracks going out to the point of the bluff, I found the den. As I stood four or five yards from the entrance, the female furtively pushed her head out of the burrow, then, on seeing me, withdrew. But in a moment she came out with a rush and galloped part way down the slope, where she stopped a moment to bark. She loped away and joined the male, and both parents howled and barked from the nearby ravine until I left.

I was sorry I had come upon the den in this sudden way, for I feared that the young would be moved and that I might fail to find the new location. From the den issued the soft whimpering of the pups. I could not make matters much worse, so I wriggled into the burrow, which was sixteen inches high and twenty-five inches wide, to investigate the young. Six feet from the entrance was a right-angle turn. Here the burrow was enlarged to form a bed which the female apparently had been using, for it was well worn. But, with the melting of the recent snowfall, this bed was full of water, in which there was a sprinkling of porcupine droppings. A porcupine had used the den the preceding winter. The willows nearby had been barked by the porcupine in its winter feeding, just as the Tanana Indian had found the spruces barked near a den.

From the turn, the burrow slanted slightly upward for six feet to the chamber in which the pups were huddled squirming. With a hooked willow I managed to pull three of the six to me. Not wishing to subject them all to even a slight wetting in the puddle at the turn, and feeling guilty about disturbing the den so much, I withdrew with the three I had. Their eyes were still closed and they appeared to be about a week old. All three were females, dark, almost black. One seemed slightly lighter than the other two, and I placed her in my packsack to take back to camp and raise for closer observation and acquaintance. We later named this wolf pup "Wags," and she was to give us many interesting hours. Then I crawled into the den again with the other two and returned them to their snug chamber.

17. Battles at the East Fork Den

IT SEEMED CERTAIN THAT, after my overly intimate intrusion
on the wolves in the East Fork River den, the family would
move. The following morning, therefore, I walked toward the
den to take up the trail before the snow melted. But from a
distance I saw the black male curled up on the point about
fifteen yards from the entrance. They apparently had not
moved away after all.

On a ridge across the river from the den, about half a mile
away or less, I found a good observation point and in the following weeks was able to watch the den unobserved. All this
area was in the open tundra beyond the trees, and I had a
view of the den and the surrounding landscape for several
miles in all directions. Between May 15 and July 7 I spent
many hours on the ridge watching the home life of this Alaska
wolf family. The longest continuous vigil was thirty-three
hours, and twice I observed them all night.

I was not always alone, however; when I returned to Alaska
the second season, my family accompanied me. My daughter
Gail, then four years old, was fascinated by my reports and
begged to spend a night on the ridge. One evening I packed
her and the sleeping bags across the many shallow channels of
the river, and we settled down to watch from the observation
point. She had trouble with the binoculars and, for the most

part, listened to my account of what was happening. But, after all, was not the adventure she had visualized chiefly to be out under the open sky, to sleep out near the wolves on the high ridge among the arctic flowers in the northern all-night twilight?

As I watched from the ridge, I learned more about the wolves. For one thing, in so far as I am aware, it had been taken for granted that a wolf family consisted of a pair of adults and the pups. Perhaps that is the rule, although we may not have enough information about wolves to know with certainty. Usually, when a den is discovered, the young are destroyed and all opportunity for making further observations is thereby lost.

On May 26, a few days after beginning an almost daily watch of the den, I was astonished at seeing two strange gray wolves move from where they had been lying a few yards from the den entrance. These two wolves proved to be males. They rested at the den most of the day. At four P.M., in company with the black father wolf, they departed for the night hunt. Because I had not watched the den closely the first week after finding it, I do not know when the two gray males first made their appearance there, but, judging from later events, it seems likely that they were there occasionally from the first.

Five days later I saw a second black wolf, a female, making a total of five adults at the den — three males and two females. These five wolves lounged at the den day after day until the family moved away. There may have been another male in the group, for I learned that a male had been inadvertently shot about two miles from the den a few days before I found it.

Late in July I saw another male with the band, and a little later a fourth extra male joined them. These seven wolves, or various combinations of them, I saw together frequently in

August and September. Five of the seven were males. The
four extra males appeared to be bachelors.

I do not know the relationship of the pair to the two extra
males and the extra female at the den. The extras may have
been pups born to the gray female in years past, or they may
have been her brothers and sister, or no blood relation at all.
I knew the gray female in 1939. She was then traveling with
two gray and two black wolves, which I did not know well
enough to be certain they were the same as those at the den
in 1940. But since the color combination of the wolves traveling together was the same in 1940 as in 1939, I am quite certain
that the same wolves were involved. Apparently, then, all the
adult wolves at the den in 1940 were at least two years old. In
1941 I knew that the extra male with the female was at least
two years old, for he was an easily identified gray male which
was at the den in 1940. The fact that none of the 1940 pups
was at the 1941 den supports the conclusion that the extra
wolves at the 1940 den were not the previous year's pups.

The presence of the five adults in the East Fork family during denning time in 1940, and three in 1941, and the presence
of three adults in the Savage River family, suggest that it may
not be uncommon to find more than two adults at a den. The
presence of exra adults is an unusual family make-up which is
probably an outcome of the close association of the wolves in
the band. It should be an advantage for the parents to have
help in hunting and feeding the pups.

Wolves vary much in color, size, contour, and action. No
doubt there is also much variation in temperament. Many are
so distinctively colored or patterned that they can be identified
from afar. I found the gray ones easier to identify, since there
is more individual variation in color pattern among them than
in the black wolves.

The mother of the pups was dark gray, almost "bluish," over the back and had light under parts, a blackish face, and a silvery mane. She was thick-bodied, short-legged, short-muzzled, and smaller than the others. She was easily recognized from afar.

The father was black, with a yellowish vertical streak behind each shoulder. From a distance he appeared coal black except for the yellow shoulder marks, but a nearer view revealed a scattering of silver and rusty hairs, especially over the shoulders and along the sides. There was an extra fullness of the neck under the chin. He seemed more solemn than the others, but perhaps I partly imagined this, knowing as I did that many of the family cares rested on his shoulders. On the hunts that I observed he usually took the lead in running down caribou calves.

The other black wolf was a slender-built, long-legged female. Her muzzle seemed exceptionally long, reminding me of the Little Red Riding Hood illustrations. Her neck was not as thick as that of the black male. This female had no young in 1940 but had her own family in 1941.

What appeared to be the largest wolf was a tall, rangy male with a long, silvery mane and a dark mantle over the back and part way down the sides. He seemed to be the lord and master of the group, although he was not mated to any of the females. The other wolves approached this one with some diffidence, usually cowering before him. He deigned to wag his tail only after the others had done so. He was also the dandy in appearance. When he trotted off for a hunt, his tail waved jauntily and there was a spring and sprightly spirit in his step. The excess energy at times gave a rocking-horse gallop quite different from that of any of the others.

The other gray male at the den I called "Grandpa" in my

notes. He was a rangy wolf of nondescript color and without distinctive markings. He moved as though he were old and a little stiff, and sometimes he had sore feet which made him limp. From all appearances he was an old animal, although in this I may be mistaken.

One of the grays that joined the group in late July was a large male. His face was light, except for a black robber's mask over the eyes which was distinctive and recognizable from a distance. His chest was conspicuously white, and he moved with much spring and energy.

The other wolf, which joined the group in August, was a huge gray animal with a light, yellowish face. In 1941 he was mated to the small black female, which had had no young the preceding year.

All these wolves could be readily distinguished within the group, but some of the less distinctively marked ones might have been confused among a group of strange wolves. The black-faced gray female, the robber-masked male, and the black-mantled male were so characteristically marked that they could be identified in a large company.

I suppose that some of the variability exhibited in these wolves could have resulted from crossings in the wild with dogs. Such crossings in the wild have been reported, and the wolf in captivity crosses readily with dogs. Years ago, at Circle, Alaska, a wolf hung around the settlement for some time, and dogs were seen with it. The people thought that the wolf was a female attracted to the dogs during the breeding period. Perhaps sufficient variability is inherent in the species, however, to account for the variations I noted in the park and in skins I have examined; the amount of crossing with dogs has probably not been great enough to alter much the genetic composition of the wolf population.

I spent many hours watching the wolves at the den, yet, when I undertake to write about it, there does not seem to be a great deal to relate. There were exciting incidents, however, such as the time a strange wolf was driven away. This happened on May 31, 1940, when all five adults were at home. Between ten A.M. and noon the mantled male had been on the alert, raising his head to look around at intervals of two or three minutes. Several times he changed his position until he was about two hundred yards above the den. Such prolonged watchfulness was unusual, but it was explained by later events. Shortly after noon the four wolves at the den joined the mantled male and they all bunched up, wagging tails and expressing much friendliness. Then I noticed a sixth wolf, a small gray animal, about fifty yards from the others. No doubt it was the presence of this wolf that had kept the mantled male so alert during the preceding two hours.

All the wolves trotted to the stranger and practically surrounded it, and for a few minutes I thought there was just the suggestion of tail wagging by some of them. But something tipped the scales the other way, for the wolves began to bite at the stranger. It rolled over on its back, begging quarter. The attack continued, however, and it scrambled to its feet and with difficulty emerged from the snapping wolves. Twice it was knocked over as it ran down the slope with the five wolves in hot pursuit. They chased after it about two hundred yards to the river bar, and the mantled male crossed the bar after it. The two ran out of my sight under the ridge from which I was watching.

Four of the wolves returned to the den, but the mantled male stopped halfway up the slope and lay down facing the bar. Presently he walked slowly forward as though stalking a marmot. Then he commenced to gallop down the slope again toward the stranger, which had returned part way up the slope. Back on the bar the stranger slowed up, waiting in a fawning attitude for the mantled male. The latter snapped at the stranger, which rolled over on its back, again begging quarter. But the stranger received no quarter, so again it had to run away. The male returned up the hill, tail held stiffly out behind, slightly raised. When he neared the den the four wolves ran out to meet him, and there was again much tail wagging and evidence of friendly feeling.

The unfortunate stranger's hip and base of tail were soaked with blood. It was completely discouraged in its attempt to join the group, for I did not see it again. It may have been forced to leave the territory of this wolf family, for if it were encountered, it probably would have been attacked again. Judging from the usual reaction of a group of dogs to a strange dog, such treatment of a strange wolf would seem normal. Small groups of wolves may be treated like this lone wolf, hence it is advantageous for minor packs to find territories where they are unmolested. Such rough treatment of individual wolves, if it is normal, would tend to limit the number of wolves on a given range.

Other interesting incidents at the den involved grizzly bears. As a rule, grizzlies and wolves occupy the same range without taking much notice of each other, but not infrequently the grizzlies discover wolf kills and unhesitatingly dispossess the wolf and assume ownership. The loss is usually not a serious matter to the wolves, for if food is scarce the kills will generally be consumed before the bears find them. In the relationship existing between the two species, the wolves are the losers and the meat-hungry bears are the gainers.

When the bears take possession of a kill in the presence of wolves, they are much harassed, but they are so powerful that the wolves must be careful to avoid their strong arms. The wolves must confine their attack to quick nips from the rear. But the bears are alert, and usually the wolves must jump away before they come near enough for even a nip.

At the East Fork wolf den were observed two bear-and-wolf encounters. The first one, which took place on June 5, I did not see, but it was reported to me by Harold Herning. A female grizzly with three lusty two-year-old cubs approached the den from downwind. They lifted their muzzles as they sniffed the enticing smell of meat, and advanced expectantly. They were not noticed until they were almost at the den, when the four adult wolves that were at home dashed out at them, attacking from all sides. The darkest cub seemed to enjoy the fight, for he would dash at the wolves with great vigor and was sometimes off by himself, waging a lone battle. (On later occasions I noticed that this cub was particularly aggressive when attacked by wolves.) The four bears remained at the den for about an hour, feeding on meat scraps and uncovering meat the wolves had buried. During all this time, the bears were under attack. When the pillaging was complete, the bears moved up the slope.

The following morning I was at the wolf den a little before eight o'clock. The female grizzly and the three cubs were on a snowbank about half a mile above the den. The cubs were inclined to wander down to the den when the bears started for the river bar, but the female held a course down a ravine to one side. On the bar they fed on roots, gradually moving out of view behind a hump of the ridge I was on.

At ten o'clock the black male wolf returned to the den, carrying food in his jaws. He was met by four adults and there was much friendly tail wagging. While the wolves were still bunched, a dark object loomed up in the east. It was a grizzly and it appeared to be following a trail, probably the trail of the female grizzly with the three cubs, for they had come along that way the day before. The bear was in a hurry, occasionally breaking into a short gallop. It is possible that this was a male interested in the female. As it came downwind from the den, it threw up its muzzle and sniffed the air, no doubt smelling both meat and wolves. It continued to gallop forward. The five wolves did not see the grizzly until it was a little more than a hundred yards away. Then they galloped toward it, the black male far in the lead. When the bear saw the approaching wolves, it turned and ran back over its trail, with the black wolf close at its heels. The bear retreated a few jumps at a time but had to turn to protect its rear from the wolves, which tried to dash in and nip. When all the wolves had caught up with the bear, they surrounded it. As it dashed at one wolf, another would drive in from behind, and then the bear would turn quickly to catch this aggressor. But the wolves were the quicker and quite easily avoided its rushes. Sometimes the lunge at a wolf was a feint, and in the sudden turn following the feint the bear would almost catch a wolf rushing in at its rear. As it lunged at a wolf, both paws reached forward in what appeared to be an attempt to grasp it. There was no quick slapping at a wolf with its powerful arms. The target was perhaps too distant for such tactics. After about ten minutes the two female wolves withdrew toward the den and shortly thereafter the wolf identified as Grandpa moved off.

The black male and the black-mantled male worried the bear for a few minutes, and then the latter lay down about seventy-five yards away. A few minutes later the black father also departed. Left alone, the bear resumed its travels in a direction which would take it a little to one side of the den; but not for long. The black-mantled male quickly attacked and the other four wolves approached at a gallop. After another five minutes of worrying the bear, the wolves moved back toward the den, the black male again being the last to leave. The bear turned and slowly retraced its steps, disappearing in a swale a half mile or more away. It did not seem that the wolves actually bit the bear. The bear did not touch any of the wolves, although once the black-mantled male escaped from the bear's outstretched arms only by strenuous efforts. On the whole, the wolves had surely discouraged the bear with their spirited attack.

Ranger Harold Herning reported seeing a grizzly appropriate a calf caribou soon after it was killed by a wolf. Two of the five wolves present attacked the bear, but after being chased a few times they retired. Having killed three other calves and eaten their fill, they probably did not have a strong desire to attack the bear.

At a road-camp garbage dump the female bear with the three two-year-old cubs often met the wolves in the common search for choice bits. Here the wolves walked about, at times, within a few yards of the bears. One evening the bear family approached the pit four abreast as the black-mantled male and the black male wolf fed. The black one moved off a few yards to one side and the other wolf looked back at the bears a few times as they came, but fed with tail toward them until they were eight or nine yards away. Then he easily avoided a charge made by one of the cubs. The two wolves maneuvered among the bears, which brought their food out of the pit to eat. Frequently the big cubs chased the wolves, but the latter easily avoided the rushes. Once a wolf walked between two bears which were only seven or eight yards apart but, in doing so he watched them closely. After a half hour of this activity, the wolves lay down to wait for the bears to depart.

On September 22, 1940, this bear family and the wolves met not far from the garbage pit. On this occasion the black male chased one of the cubs for a short distance, then the cub turned and chased the wolf. Variations of this were repeated several times.

These particular bears and wolves had more frequent contact than usual because of the road-camp garbage pit which attracted them. The bears were seen robbing the wolf den only once.

18. Wolf Home Life

OFTEN, as I watched the den from my observation point on the ridge, there was little activity. But every hour spent on the lookout was charged with expectations; and the presence of the wolves created an aura over this wilderness.

Just as a laboring husband comes home to the family each evening after working all day, so does the wolf come home each morning after working all night. The wolf comes home tired, too, for he has traveled far in his hunting. Ten or fifteen miles is a usual jaunt, and he generally takes part in some chases in which he exerts himself tremendously. His travels take him up and down many slopes and ridges. When he arrives at the den he flops, relaxes completely, and may not even change his position for three or four hours. Often he will not even raise his head to look around for intruders. Sometimes he will stretch and yawn, change his position, or shift his bed a few yards. In summer he usually lies stretched out on his side but once in a while will be curled up as in winter. Frequently a wolf will move over to a neighbor, perhaps sniff of him, getting for response only a lazy indication of recognition by an up-and-down wag of the tail, and lie down near him.

An animal might move from the point of the bluff down to the gravel bar, or, while the overflow ice still remained on the bars, he might lie in the snow for a while. When the caribou grazed near the den, a wolf might raise himself up a bit for a look, but generally a caribou was not sufficient reason for him to disturb his resting. The female might be inside the den, or on the outside, for hours at a time. The five adults might be sleeping a few hundred yards apart, or three or four of them within a few yards of each other. Of course, not all the adults were always at home; one or two might be out for a short daylight excursion or fail to come home after the night hunt. That, in brief, was the routine activity at the den.

For the first few weeks the gray female spent much time in the den with the pups, both day and night. When she was outside she usually lay only a few yards from the entrance, although she sometimes wandered off as far as half a mile to feed on cached meat. When the rest of the band was off on the night hunt, she remained at home, except on three occasions that I know of — June 1, 8, and 16. Each time she went off with the band she ran as though she were in high spirits, seeming happy to be off on an expedition with the others. On these three occasions the black female remained through the night with the pups.

The father and the black female entered the den when the pups were only a couple of weeks old. Later, when the pups were old enough to toddle about outside, the father and the two females were very attentive to them. The two gray males often sniffed at the pups, which frequently crawled over all five wolves in their play. Sometimes the pups played so much around an adult that it would move away to a safe distance where it could rest in greater peace.

The attentiveness of the black female to the pups was remarkable. It seemed at times that she might have produced some of them, and I do not absolutely know that she did not. But her absence from the den the first ten days (so far as I know), the uniformity in the size of the pups, and the greater concern and responsibility exhibited by the gray female strongly indicate that the gray one had produced all the pups. The companionship of these two adults suggests that two females might at times den together, although their having pups in one den would be somewhat inconvenient. Rather, one would expect them to den near each other as these two females did in 1941.

Wolves have few enemies and consequently are not often watchful at the den or elsewhere. I sometimes approached surprisingly close to the wolf band before being discovered. Several times I was practically in the midst of the band before I was noticed. Once, after all the others had run off, one wolf which must have been sound asleep got up behind me and, in following the others, passed me at a distance of only about thirty yards. These wolves were scarcely molested during the course of the study, so they may have been less watchful than in places where they are hunted. But their actions were probably normal for primitive conditions. When alert, their keen eyes do not miss much.

Before the vegetation changed from brown to green, the gray wolves, when curled up or when only the back showed, were especially difficult to see against the brown background. But, if they were stretched out so as to expose the light under parts, they were plainly visible. The black ones were usually more conspicuous, but under certain conditions of poor light or dark background the gray wolves were the more conspicuous. At the den, it was sometimes difficult to see all the wolves because of slight depressions in which they lay, and the hummocks hiding them. Once, when all five adults were lying on the open tundra slope above the den, not one could be seen from my lookout. Often only two or three of the five could be seen until some movement showed the position of the others.

The strongest impression remaining with me after watching the wolves on numerous occasions was their friendliness. The adults were friendly toward each other and amiable toward the pups, at least as late as October. This innate good feeling has been strongly marked in the three captive wolves which I have known. Undoubtedly, however, wolves sometimes have their quarrels.

It is likely that all the wolves brought food to the East Fork den. It was necessary to bring food for the pups and for the female remaining with them. I observed the gray female, the black male, and the mantled male carrying food. Apparently, much of the food was eaten at the kill and regurgitated at the den for the pups, but I was not close enough to observe any regurgitation.

Wolves, in common with most flesh eaters, often cache excess food for future use. Once I noted a sheep horn which had been carried some distance and buried in the snow; another time the head of a ram was hidden in the snow; again, two pieces of sheep meat were cached, only to be consumed shortly by foxes. But the food is not always cached. When there is an abundant supply, the bother of caching the food is often omitted. I have found calf caribou on the calving grounds left untouched where killed. The wolves were seemingly aware that there was not much point in caching them, since food was readily available on all sides.

I observed a good example of provident caching on July 19, 1941, after a wolf had killed a caribou calf on the bars of the East Fork River. At this time the caribou herds had moved out of the region, and since food was not readily available, it was worth caching. The wolf was hungry, for she ate voraciously for more than half an hour. Three times during the meal she walked to the stream for a drink. After feeding she got my scent, circled above me, barking and howling, then retreated toward the den, still not having seen me. I waited for almost an hour before I saw her coming down the river bar along the opposite shore. She trotted directly to the carcass and, after feeding on a few morsels, she chewed until a foreleg and shoulder had been severed. With this piece in her jaws she waded the stream and trotted about three hundred yards up the bar. Here she stopped and, still holding the leg in her mouth, pawed a shallow hole in the gravel and placed the leg in it. Then, with a long sweeping motion of her head she used her nose to push gravel over the leg. The job was quickly completed and she trotted back to the carcass, chewed off the head and, this time, buried it about three hundred yards away without crossing the stream. On the third trip she carried another leg and cached it on the side of the river from which I was watching. When she returned to the carcass from this trip, the wind shifted, bringing my scent to her. Without hesitating, she trotted briskly across the stream and up the bar, not stopping until at least half a mile separated us.

Many of the caches made by wolves are utilized by bears, foxes, eagles, and other flesh eaters. These others probably use the caches about as much as do the owners.

At the East Fork den, some of the food was brought directly to the den, where I often saw the young feeding on it, but much of it was cached one or two hundred yards away, and some of it as much as half a mile away. The wolf remaining at home during the night went out to these food caches, and occasionally one of the other wolves might eat a little from them during the day. The wolves that hunted probably ate their fill near the kill.

Relatively few bone remains were to be found at any of the dens. At the Lower Toklat den there was the skin of a marmot, four calf caribou legs, leg bones of an adult caribou, and three hundred yards away the head of a cow caribou. Remains, mainly hair, of an adult caribou were found on the bar about a quarter of a mile from this den. At the East Fork den there were scarcely any bones around the den after the family left, and only a few where they lived after leaving the den.

There was considerable variation in the time of departure for the night hunt. On a few occasions the wolves left as early as four P.M., and again they had not left at nine or nine-thirty P.M. I saw them departing for the hunt eleven times: five of these times they left between four and five forty-five P.M. and six times they left between seven and nine-thirty P.M. Usually the hunting group consisted of the three males, but sometimes one of the females was in the group. The wolves hunted in a variety of combinations — singly, in pairs, or all together. In the fall the adults and young traveled together much of the time, forming a pack of seven adults and five pups.

Usually the wolves returned to the den each morning, but three wolves which left the den at four P.M. on May 26 had not returned to the den the following day by eight P.M., when I left the lookout after watching all night and day. The wolves had probably spent the day near the scene of their hunt. These wolves were back at the den on May 28.

Considerable ceremony often precedes the departure for the hunt. Usually there is a general get-together and much tail wagging. On May 31 I left the lookout at eight-thirty P.M., since the wolves seemed, after some indications of departure, to have settled down again. But as I looked back from the river bar on my way to camp, I saw the two blacks and the two gray males assembled on the skyline, wagging their tails and frisk-

Considerable ceremony often precedes the departure for the hunt. Usually there **is** a general get-together and much tail-wagging.

ing together. There they all howled, and while they howled, the gray female galloped up from the den a hundred yards away and joined them. She was greeted with energetic tail wagging and general good feeling; the vigorous actions came to an end, and five muzzles pointed skyward. Their howling floated softly across the tundra. Then abruptly the assemblage broke up. The mother returned to the den to assume her vigil and four wolves trotted eastward into the dusk.

On June 2 some restlessness was evident among the wolves at three-fifty P.M. The two gray males and the black male approached the den, where the black female and some pups were lying. Then the black male lay down near the den; the mantled male walked down on the flat a hundred yards away and lay down, and Grandpa, following him, continued along the bar another 150 yards before he lay down. At six forty-five P.M. the mantled male sat on his haunches, howled three times, and in a few minutes sent forth two more long, mournful howls. Grandpa stood up and with the mantled male trotted a few steps toward five passing caribou. Then the mantled male howled six or seven times, twice while lying down. The gray female trotted to the gray males, and the three of them stood together, wagging their tails in the most friendly fashion. The mantled one howled and they started up the slope. But before going more than two hundred yards they lay down again. A few minutes later, at seven-fifteen P.M., the mantled male howled a few times and walked to the den, followed by Grandpa. The latter seemed ready to go whenever anyone decided to be on the move. At the den the black female squirmed and crouched before the mantled male, placing both her paws around his neck as she crouched in front of him. This hugging with the front paws is not an uncommon action.

Later the two gray males and both black wolves were in a huddle near the den entrance, vigorously wagging their tails and pressing against each other. The gray female joined them from up the slope; the tail wagging became more vigorous and there was a renewed activity of friendliness. At seven-thirty P.M. the mantled male descended the slope to the bar and started to trot away. He was shortly followed by the black male and Grandpa. The black female followed the departing males to the bar, then returned to the gray female at the den. Both females remained at the den this time.

On June 8, at seven-fifteen P.M., Grandpa approached the mantled male, wagging his tail. The mantled male stood stiffly erect and wagged his tail slowly, with a show of dignity. The two walked over to one of the blacks and lay down. The mantled male turned twice around before lying down, like a dog, then rose and turned around again before settling down.

There was no movement until nine o'clock in the evening, when Grandpa rose, shook himself, and walked over to the mantled male. They wagged tails and were joined in the ceremony by the black female. The mantled male sniffed at the black male, which was still resting. He rose and the tail wagging began again. The gray female hurried down the slope to the others and the tail wagging became increasingly vigorous. The friendly display lasted seven or eight minutes, and they started eastward, led by Grandpa, who seemed especially spry this evening. The black female followed a short distance, then stopped and watched them move away. A quarter of a mile farther on, the four wolves commenced to play on the green flat. The black female trotted rapidly to them and joined in the play. After a few minutes of pushing and hugging, the four again started off, this time abreast, spaced about fifty to seventy-five yards apart. The black female followed for a short distance and lay down. She appeared anxious to follow. After fifteen minutes she returned to the den, and two or three pups came out to join her.

The gray female at first led the wolves up the long slope toward Sable Pass, but later the two gray males were in front, running parallel about two hundred yards apart. They trotted most of the time, but galloped up some of the steeper slopes. On a snow field they stopped for a time to frolic. The black female remained at the den all night. The hunters returned at nine-fifteen the following morning. The gray female hurried unhesitatingly to the burrow, like a mother who has been absent from her child for a few hours. The black male flopped over on his side a short distance from the den and lay perfectly still and relaxed. About one mile north of the den, on a high point, the mantled male was stretched out on his side. The wolves had been away on the hunt about twelve hours.

At four P.M. on June 16 the two gray males, the black male, and the gray female left the den, led by the mantled male. Soon the female took the lead and headed for a spot where some eagles were feeding. She nearly captured one of the eagles by jumping high in the air after it as it took off. These wolves went directly to Teklanika River, some seven or eight miles from the den.

It was evident that by evening the wolves were rested and anxious to be off for the night's hunting. The time of their departure for the hunt no doubt varied from day to day, depending somewhat upon how soon they came in from the previous night's expedition. Theirs is not a lazy life, for the nature of their food demands that they travel long distances and work hard for it, but they seem to enjoy their nightly excursions.

The extended wanderings of the pups below the den on the river bar early in July indicated that the family might soon move away. I had refrained from approaching close enough to take pictures, but with the departure seemingly imminent, I made a careful stalk to the bank opposite the den on July 8. I was a day late, for, although I waited until evening, not a wolf was seen.

In the evening I looked over at the den from Polychrome Pass. At eight o'clock the mantled male came from the direction of Sable Pass. He stopped several times near the den and appeared to be howling, but he was too far away for me to hear him. He moved to a knoll south of the den and sniffed about in a short ravine where the pups had often been seen. Apparently the family had moved during his absence. He dropped down to the bar and followed southward along the bank for half a mile, then abruptly turned and climbed the bank at a gallop. Above, on the sloping tundra, he joined the female and the pups and for a time they wagged tails and romped together. The days at the den were over.

The following morning, July 9, I walked up the river past the wolf den and across the broad level bar covered with grass and mountain avens toward the spot where I had seen the wolf family the previous evening. The gray female and both black wolves were with the pups and saw me when I came in view around the point on which the den is located. They sat on their haunches watching me approach. There had been no chance to make a stalk, so I continued forward, hoping that the wolves would stay close enough to the pups to permit me to take some pictures. After watching me advance for about two hundred yards, the three adults ran up the long, open slope, stopping at intervals to bark and howl. The black male, after angling up the slope, galloped along the hill in my direction, keeping his elevation above me and frequently stopping to bark. I continued forward and passed the three wolves, which now were barking at me from directly up the slope. The gray female joined the black male, but the black female moved higher up. When I was almost opposite and within about a quarter of a mile of the pups (they had taken refuge in a burrow ten feet long and open at both ends), the black male galloped down the slope to the bar, followed closely by the gray female. They came out on my trail and headed directly into the wind toward me at a gallop. The female took the lead and, with noses to the ground, they came on at a brisk trot.

I set up the movie camera and saw them in the finder, running silently and swiftly. Their purposefulness and intent manner worried me some, and I began to wonder if they would turn aside. They were accustomed to seeing people and lacked the timidity of most wolves. I wondered if the two grays and the other black might not join the two coming toward me. Generally, I carried an automatic pistol in my packsack, and as I had not checked on the matter before starting out, I now hurriedly looked and was relieved to find it was there. By that time the wolves were about a hundred yards away and, circling to one side, they commenced to bark. The female passed me and the male remained on the other side. Both continued howling and barking, now about two hundred yards away. After exposing my film, I walked down the bar. The female remained opposite the pups, howling at intervals, and the male kept abreast of me for half a mile as I went down the bar to camp. The black female remained on the slope, howling. When I returned to the spot an hour and a half later with more film, the wolves had all departed. I did not see the pups again until August 22, when I found them about five miles away. I saw the adults often, but the pups were not traveling with them. They were apparently at some rendezvous.

At six A.M. on July 30 I heard deep howling a short distance above our camp. With my camera I hurried toward the sound and came upon the mantled male on the flat below Sable Mountain. Presently, he was joined by the gray female and Grandpa. They howled together and were answered by a wolf farther up the gentle slope. The three wolves moved nearer the base of Sable Mountain, where they joined the black-masked wolf. They lay on the tundra, in a depression just sufficiently deep to hide them from me.

Later, the black male came from the west and joined them. His coming was heralded by the loud chirping of the ground squirrels all along his route. He walked slowly to the mantled male and was surrounded by the other wolves, all wagging their tails. The black male walked about thirty yards away and they all lay down again. Later they stood up and after some more friendly tail wagging lay down.

Early in the afternoon Ranger Harold Herning and I advanced cautiously toward the band for pictures, taking advantage of the shallow swale below the animals. We were about a hundred yards from them when we noticed the black male peering over the rise and saw him trot to one side and watch us. The others, who had not seen us, trotted over to the black male. There was a slight altercation, accompanied by a little growling and snarling, when they came together. Grandpa lay down, but about that time all the wolves saw us. They watched us for a few seconds before trotting up the slope, still not much afraid. They continued over a rise and disappeared up the mouth of a short canyon. When we came in view again, they were moving slowly up the canyon. Near its head they laboriously climbed a steep rock slope, using a switchback technique. On top they followed the ridge along the sky line for some time before disappearing.

The black female was absent from the group; she probably was with the pups. The band was resting up for the night hunt and may not have been far from the pups. From this time on the black-masked male was frequently with the band.

On August 22, on the flat below Polychrome Pass, I saw the pups for the first time since they left the den on July 9. The wolves seemed to have been attracted to this vicinity by the refuse from a nearby road camp. The men at the road camp had heard the wolves howling for several nights, so the family probably had been there for a few days before I saw them.

At four P.M. a wolf howled three times from a point southeast of the refuse heap, and an hour later the gray female, followed by a pup, appeared from behind a bench. She apparently was on her way to the refuse heap. Out on the flat the mantled male walked slowly into view from behind the same bench, followed by two black pups a hundred yards to the rear. He walked slowly, with head down and tail held horizontally. Some distance to the west the two adults met and moved westward. The pups did not follow but returned to the east and lay down in sight of us.

The following day, with two companions, I returned to

Polychrome Pass and saw the wolves lying on the tundra among the dwarf birches and short willows.

As we watched, a large gray wolf with a light face walked toward the others. He looked over the flat where the wolves rested and lay down on the beach a short distance from them. This gray wolf was the second addition, and there were now seven adults in the band.

We stalked the wolves, coming first to the large gray one on the bench. He rose fifty yards or so ahead of us and loped away toward the others on the flat and aroused them. They ran off from in front of us, all headed southward. I hurried over the flat to get a picture of a black pup which was standing uncertainly, watching the others run. While I was photographing the pup, the mantled male got up behind me a short distance and ran close past me between me and my companions. He must have been sound asleep to be aroused so tardily. Five of the adults and some of the pups stopped on a knoll about a mile away. The parents hung back, barking at us, probably solicitous over some of the pups which had been left a distance behind. When I walked toward them, they barked and howled and those on the knoll howled in the usual mournful chorus. Soon the parents hurried to join the others. The pups in the rear must have caught up with the band by this time. The green grass at the base of a bench near the place where the wolves had been lying was flattened and worn, showing that the wolves had spent much time there. I noted the bones of a fresh front leg of a large caribou nearby and found several droppings.

On September 17 I saw the family again on the flat below Polychrome Pass. At eight A.M. three pups trotted briskly down a gravel bar. They stopped at one spot and sniffed zealously, apparently in search of morsels where they had feasted

during the night. Nothing was there except rich sniffing. They
climbed a bluff above the bar, scared an eagle from its perch,
and sniffed about on a point of rock. They returned to the bar
and hurried to the gray female and the black-masked male,
which were lying within a few feet of each other. The mantled
male had just lain down on a bench above these two and was
hidden in the dwarf birch. The three black pups, now almost
the size of their mother, swarmed all over her and later touched
noses with the black-masked male, which had joined the band
after the pups left the den.

The mother led the way to the base of a bench and there uncovered some morsels of food which were at once eaten by the
pups. They resumed their play all over the mother, after
which they all lay down near the black-masked male, forming
a circle. Another black pup that had been off by itself hunting
mice approached the group and smelled of the black-masked
male; the adult sniffed at the pup in turn, causing it to roll
over on its back with diffidence. The pup then smelled of each
of the others; they barely acknowledged the salutation by raising their noses a trifle. A yellow pup that had been hunting
mice also joined the group. It first smelled noses with the
black-masked male, which raised his nose to it and, as he lay
flat on his side, wagged his tail a few times. Then the pup, wagging its tail all the while, sniffed noses with each of the other
pups, which were stretched out flat. The mother now trotted a
mile to the east on some errand and returned a couple of hours
later.

In the afternoon, some of the pups hunted mice. At four P.M.
they all moved a mile out on the flat and lay down. An hour
later Grandpa showed up, sniffed around where the wolves
had been resting, and continued southward on their trail.

I saw the wolf family at Polychrome Pass on September 22,

23, and 24. But on September 28 they had moved to a point on Teklanika River, twenty miles away. My attention was first attracted by the yellow pup, which disappeared in a fringe of trees. Later I heard the howling of several wolves and saw the yellow pup trot in the direction of the sound and join the four black pups. Soon they all galloped out of sight. I advanced cautiously and came upon the five pups, their parents, and Grandpa, 140 yards from me. I exposed some motion-picture film, then dropped out of view to change film. While I was thus occupied, they all howled, and there was considerable barking, which resembled the yapping of coyotes. When I again peered over the rise, all but the black male were moving away with much tail wagging and milling around. The black male saw me and trotted after the others, and all disappeared around the base of a low ridge. On my way back to the road I met the other four adults heading toward the spot where the wolves had howled. Apparently they had heard the noise too. The mantled male was quite surprised when he saw me 150 yards away and made several high jumps with his head turned toward me. They all stopped to watch me, then slowly trotted on around the ridge after the others. This was about nine-thirty A.M.; at three o'clock I found all the wolves resting near the base of a long slope about a mile away. They saw me approach in the distance and moved a short way up the slope, from which they watched me. The following day I saw the band four miles to the north, but I was unable to stalk them.

Although I saw tracks, presumably of the East Fork family, during the winter, the wolves themselves escaped observation until March 17, when I came unexpectedly upon the band at Savage River, about thirty miles from the East Fork den site. I followed several fresh tracks which crossed my way and led out onto an open flat where there were many bare spots made by the caribou in pawing for grass and lichens. One of the many spots appeared a little different from the others; looking at it through field glasses, I saw that it was a black wolf stretched out on its side. Searching the flat ahead of me, I made out two more black wolves. Off to one side of these a gray wolf sat up and, before curling up, looked about at random without noticing me as I crouched about three hundred yards away. To reach a strip of woods from which I could watch the wolves unobserved, I backtracked cautiously but was discovered just as I was about to enter the woods. The black wolf that saw me aroused all the others when it howled. At least ten wolves came to life and after a brief view of me hurried away, kicking up sparkling puffs of snow as they galloped.

I have few data on the dispersal of the young. The 1940 East Fork pups remained with the pack through the first winter until at least March 17, 1941, the last date I saw the pack together. On May 14 I saw one of the pups two miles from the den, and three days later I saw another at a similar distance from the old home. This was the last time I saw the 1940 pups. I never saw any of these young at the den in 1941. Some may have been trapped during the winter, but at least three or four escaped the trappers.

The East Fork den, which was used in 1940, was again used in 1941. As in the previous season, the black male was mated to the gray female. On my first visit to the area on May 12, I saw the black male lying close to the den entrance. The mantled male headquartered at the den as he had done the previous year, but I did not see Grandpa and the black-masked male all summer. Possibly they had been trapped during the winter outside the park. On June 21 four pups played on the bar and climbed over the father and the mantled male.

The black female which had helped the gray female take

care of her young in 1940 was not at the old East Fork den early in the summer, but I noted her several times in the region. On June 1 she and a large, light-faced gray male that was with the band in the late summer of 1940 were traveling together on Igloo Creek. I later learned that she was mated with this male, but I did not find her den.

On June 30 two hikers saw the black female coming up East Fork River, followed by a pup which, in crossing the river, was carried downstream some distance and treated a little roughly by the fast water. The hikers were able to run it down and catch it by the tail. They said that the mother barked at them from a point about 150 yards away. When they released the pup, the mother continued up the river toward the den occupied by the gray female.

The following day I saw the black female coming down the bar, but before I could take cover she had seen me. Instead of following down the bar, she climbed a high ridge opposite me and then dropped down on the bar below me. I hoped she was on her way to her den, but if she was, she changed her mind. After trotting down the bar a third of a mile, she climbed the bank, smelled about a knoll, then came back and climbed over a high ridge. I searched for her den but did not find it. After reviewing the places where I had seen this wolf and tracks during the summer, I felt certain that she had denned about four miles below the East Fork den occupied by the gray female.

On June 30 the black female and her pups were living at the den of her neighbor, the gray female. On July 9 the gray female and her mate had moved to a rendezvous a third of a mile above the den, where the pups spent much time in an extensive growth of willows and the parents rested on the open bar nearby. On this day the black female was still using the gray female's den. On July 12 both families were together at the rendezvous. There were ten pups — six in one litter and four larger ones in the other. Gray and black pups were present in both litters.

On July 31 a companion and I, with much crawling, managed to gain a gully adjacent to the flat where the wolves rested and played. We watched for some time from the creek, which was bordered by tall willow brush. At about one P.M. we had planned to make a close approach on one of the wolves for pictures, only waiting for the sun to shine before starting. While we were waiting, the gray female appeared on the creek bank twenty-five feet away, passed behind some willows, and reappeared fifteen feet away. She then saw us and bounded away. All the wolves were alarmed, and three adults and ten pups scattered over the bar. The general retreat was southward, but the pups crisscrossed so much that the bar seemed covered with wolves. The following day the two families were again at the rendezvous.

The East Fork wolves were known to move readily over a range at least fifty miles across. During the denning period their movements radiated from the den, and ordinarily the wolves traveled a dozen or more miles from it. But they readily traveled greater distances. In the spring of 1941, when a band of five or six thousand caribou calved some twenty miles away, the wolves traveled this distance nightly to prey on the calves.

In the winter of 1940-1941 there were many caribou along the north boundary of the park, which attracted the East Fork wolves. The hub of their movements was shifted in that direction, but they continued at intervals to make trips back through the part of the range which they used most in summer. These trips were probably made in search of mountain sheep, for there were no caribou along the route. Wolves seem to enjoy traveling, and these excursions may, in part, have been made because of their habit of being on the move. Wolves are often reported to have a circular route which they take a few times during the month, but the East Fork wolves on a number of occasions traveled both ways on the same route.

A family at Savage River lived the year round in that locality. They ranged westward ten miles to Sanctuary River, which was normally the east boundary of the range of the East Fork wolves. I do not know how far to the east these Savage River wolves wandered, but possibly they went twelve miles to the Nenana River or farther. The east-west breadth of their range appeared to be less than that of the East Fork wolves, but the former may have wandered farther in a north-and-south direction. The center of the Savage River range was some thirty-five miles east of the East Fork den.

The Toklat River den was about twenty-two miles northwest of the East Fork den. While I learned nothing of the range of this family, it probably overlapped the range used by the East Fork wolves. Wolves lived the year round at Wonder Lake, about forty miles west of East Fork.

The various wolf families seemed to have rather definite year-round home ranges which overlapped somewhat. It was significant that, within the large range of the East Fork wolves, almost every wolf seen was recognizable as belonging to that band.

19. Dall Sheep

I HAVE OFTEN CLIMBED the ridges along East Fork River in McKinley Park to find Dall sheep rams. Going over the first ridge, one enters a fairyland of green slopes. There are fragile yellow arctic poppies, many kinds of saxifrage (a delicate flower whose name means rock busters), monks-hood, the ever-present mountain avens, spring beauties, and many others, each in its chosen niche. What always impressed me was the sheer greenness of the numerous parallel ridges. Also in the high sheep country are the rosy finches, snowbirds flashing black and white, the wheatear that spends its winters in Asia, and the surfbird whose nest has been found only once, and that one in McKinley Park. And, soaring over the sheep hills or sailing swift and low over the ridge tops in search of ground squirrels, is the golden eagle, whose nest is placed on ledges among cliffs. No sheep range is complete without the golden eagle.

Always there are many rams in these lofty pastures, many old veterans with long, gracefully curved horns. There is something entrancing about a mountain-sheep horn, something about its sweep that satisfies our sense of smoothness while the ruggedness of its surface gives it character. The horns of these white sheep are especially free in their sweep, are relatively slender, and often have a pleasing amber hue. The rams feed on the gentler slopes, then retire to the terraced ledges which, here and there, break the green. Or they might lie on the smooth green slopes or on one of the numerous knolls on the ridges. A few ewes and lambs are also present, but this is chiefly a ram rendezvous in summer. The ewes may be recognized by their horns, which are only six or eight inches long and but slightly curved. As the sheep rest on prominent lookouts, their keen eyes are always alert. Apparently the nose is little used for detecting danger — perhaps because of the changing and unreliable air currents in their mountain home.

From 1906 to 1923 (the year in which I first visited the park) the sheep population was consistently high, although there was considerable sheep hunting in the area up to 1920. An old-timer told me that in 1915 and 1916, when he had hunted sheep in the area, they were very abundant and one winter he had sold forty-two sheep carcasses. At this time, several other market hunters operated in the region. Many sheep apparently were killed for the market, and many were fed to the sled dogs used in hauling the meat. At an old crumbling cabin down East Fork River, I found many old ram skulls, most of which were heaped in a pile. There were 142 horns, so at least seventy-one rams had been brought to this camp. The skulls had been split open, probably to make the brains readily available to the dogs.

From 1923 to 1928 the sheep population steadily increased. Some think the peak was reached in 1928, when, from all the information that I can gather, there were at least five thousand sheep in the park. There had been no severe winter during the time the sheep herd built up, although some sheep apparently were adversely affected by winter conditions at times. But the snow conditions in the winter of 1928-1929, according to reports, caused quite a large loss among the sheep. The Park Superintendent's report for March, 1929, describes conditions as follows: "The winter has been a hard one on sheep with the deep snow and storms. They have been driven down from the ridges and into the deep snow of the flats in their effort to get feed. They were even noticed out on the flats near the north boundary four miles from the range."

One ranger wrote me that the wolves had killed many sheep that winter but that "the big jolt" had come in April, when heavy snows covered the food. For the part of the range he had investigated near headquarters, he estimated that a third of the sheep population had died. He said that those sheep that perished were mostly the old and the yearlings.

In the spring of 1929 there was a fairly good lamb crop. The following two winters were not severe, but heavy wolf predation on the large sheep population was reported.

The most serious reduction among the sheep took place, it seems, in the winter of 1931-1932, which was much more severe for the sheep than that of 1928-1929. The Park Superintendent's report for December, 1931, states that the rangers were experiencing difficulty in making patrols because of the heavy snows and that "from all indications the sheep are going to have a hard time finding forage this winter." The January, 1932, report states that the month was very cold, that the sheep were all in good condition, but that the late snows had driven them well up toward the summits of the mountains. In February, 1932, it was reported that all records for snowfall had been broken, that seventy-two inches had fallen in six days, and that the winter of 1931-1932 would be remembered as the "year of the big snow." The Superintendent reported:

During the heavy snowfall which came on the 3d of the month, I was alone at headquarters and taxed to the utmost in shoveling snow from the roofs of the buildings. It was thought for a while that several would go down, as the snow was four feet deep in

240 *A Naturalist in Alaska*

places. I called up on the phone to get a man from the station to come up and help me out. He left there at eight A.M. and arrived here at three-ten P.M. It took him just seven hours to go the two miles, notwithstanding the fact that he had on a good pair of snow-shoes. The heavy snows that came during the fore part of the month were followed by two days of rain, then below-zero weather, and a heavy crust was formed which has caused untold suffering amongst the wild animals of the interior of the park. This is espe-cially true concerning the moose. Their legs from the knee down are worn to the bone, and each moose trail is covered with blood. It is possible to walk right up on a moose as they have not the courage or strength to run away.

For the period from 1933 to 1939, variations in sheep popu-lations are not well known. One ranger thought the ebb was reached about 1935 or 1936. In 1939 there was an excellent lamb crop and a good survival of the yearlings of the previous year. Another ranger said he thought there were more sheep in 1939 than there had been for four or five years.

The yearling losses were heavier in the winter of 1939-1940, and in the spring of 1940 the lamb crop was far below par. These losses and the small lamb crop were a definite setback to the population. The lamb crop in 1941 was excellent, but there were very few yearlings because of the few lambs the year before. The population was estimated to be between one thousand and fifteen hundred sheep, a figure I now think was a little low.

In 1945, after an absence of four years, I returned to the park for a month to check on the sheep. It was obvious that there had been a sharp decline in the population since I left in 1941. The explanation? In 1941 I observed the predominance of old sheep; this showed up especially in the rams, where age differences in the horns are much more accentuated than in the ewes. In 1945 a change in age groups had taken place. There were more youngish animals on the range and proportionately fewer old ones. A well-versed hunting guide had made a similar observation in 1946 in the Wood River sheep hills some miles east of the park. It seems probable, and was so reported, that in the big die-off of 1932, it was the younger animals that survived — those two, three, and four years old. This age group was old in 1941 and, in the next couple of years, reached the end of its life span. In the intervening years, lamb survival for replacement had been low. Hence the large decrease in the population observed in 1945. My 1945 estimate of about five hundred sheep must have been fairly accurate, because a rather thorough count in 1947 was 595. After 1947 a steady increase of the sheep was noted: the 1949 count was 795; the 1951 count was 1060; the 1953 count was 1,368; the 1955 count 1,586. A count of 1,753 sheep in 1959 indicates that the present sheep population numbers about two thousand.

At the low point in the population the law of diminishing returns no doubt operated for the wolves in their sheep hunting. Not only were fewer sheep available, but the sheep also were in the more vigorous age groups and were occupying the rougher, safer cliffs. If the wolves can be perpetuated in the park, perhaps some kind of natural, moderately fluctuating balance in the sheep population will result, so that range conditions can also be kept from deteriorating. But we must keep in mind, in considering the relationships, that even under the most primitive conditions there may have been violent fluctuations in the sheep numbers, at least suggestive of what we find in the snowshoe rabbit. Therefore we should not rush into efforts to maintain the population at some medial number.

In winter the sheep are distributed over the lower ridges, where strong winds blow many slopes free of snow and where much willow browse is also available.

When the winter snows melt sufficiently to permit freedom of travel, the range of the sheep expands greatly. Many then move nearer the main Alaska Range, some going to the heads of glacial streams. (In midwinter the snow is too deep in these regions to permit their use.) Not all the sheep make these movements, for during the summer some may be found over much of the winter range.

I had many opportunities to observe the interesting habits of the sheep while they were migrating to and from the purely summer ranges. Before venturing to cross over low intervening country, the sheep may spend hours looking over the region from the slopes, apparently to be sure the coast is clear. Sometimes they spend a day or two watching before making the attempt. Often ewes and rams move across together in a compact band.

The sheep at Sanctuary and Teklanika Canyons have three or four miles of low, rolling country to cross to reach the hills adjoining Double Mountain, their immediate destination. Three or four well-defined migration trails lead across the low country.

On June 7, 1940, a band of about sixty-four sheep, both ewes and rams, crossed from Sanctuary Canyon to the low hills adjoining Double Mountain. They started crossing about two P.M. and did not arrive at the hills until five-thirty. Most of the way the sheep traveled in a compact group, stopping frequently to look ahead. Through tall willows and scattered spruce woods they walked in single file. Just before reaching the first hills, they fed for about forty-five minutes on the flats at their base. They probably were hungry and came upon some choice food. When they emerged from the woods to the open hills, they were strung out considerably and galloped up the slope in high spirits, seeming relieved to have made the crossing.

On July 27, 1939, eight sheep made a belated migration from Sanctuary Canyon to the hills north of Double Mountain. From these hills I saw a group consisting of two old ewes, two lambs, and two yearlings, one two-year-old ram, and a young ewe coming across the flats, alternately galloping and trotting, and occasionally stopping briefly to look ahead and behind. I lost sight of them as they approached the hills to one side of me, but presently I saw them cross a creek and climb a long slope to the south. They disappeared over the top, but in a few minutes all eight sheep reappeared in precipitate flight and galloped down the slope with miraculous speed and abandon. The lambs kept up with the others. They passed within thirty feet of the place where I crouched in the willows, and I could see clearly how hard they were panting. They climbed a slope north of me and stood above some cliffs, surveying the terrain they had crossed. No pursuer was in sight, but it appeared they had been badly frightened. Later they regained their composure and fed and rested.

When sheep cross a valley, they often follow an old rocky stream bed, even though the travel would be much easier on the sod beside the stream. They probably have an instinctive feeling of safety when traveling among boulders even on level terrain, since such terrain resembles the cliffs on which they are safe. This instinct for seeking boulder-strewn country on the level would serve a good purpose if such areas were extensive enough so that an enemy could not follow on firm, smooth sod close by, as they can along the stream beds. On one occasion two ewes which were captured by wolves might have escaped if they had not followed a rocky stream while the wolves ran on the firm sod alongside the rocks.

In migrating along ridges where there are cliffs, the sheep move leisurely, frequently stopping in places for a day or longer to feed. Sometimes they may remain on a mountain along the way for a week or more. A band of about a hundred ewes and lambs fed on Igloo Mountain for a week before continuing their migration. They moved slowly around the mountain in their feeding, some days going only three hundred or four hundred yards. Sometimes a band will retrace its steps half a mile or so before going forward again. The movements in spring and fall are similar, although the fall migration may at times be hurried a little by snow. However, the sheep usually begin their fall movements before the coming of heavy snows. Single animals, or bands containing up to a hundred, may be seen in migration.

The causes of migration are difficult to determine. Vegetation influences migration to the extent that it must be satisfactory on the range sought by the sheep. Grazing animals will often follow the snow line in spring as though they were seeking the tender new vegetation. This may be partly a cause-and-effect relationship, but, at least in part, it may be purely an incidental correlation; in following the snow line, the animals may simply be driven by an urge to return to a remembered summer habitat, and not necessarily by the seasonal stage of the plant growth.

We may assume that migratory habits had a beginning in the early history of a species. Sheep may have remained on the remote summer ranges until driven out by snow. Then, after such a movement had become habitual and more or less punctual, it is reasonable to assume that their fall migration anticipated the coming of the snow. Or the inception of the fall movement may be due to the deterioration of vegetation (by ripening and drying) on the purely summer range, for the sheep usually begin to move before the snow drives them out.

Insects are often stressed as a factor in the migration of big game, and no doubt they do at times influence local movements. But they cannot be regarded as a major factor. Occasionally flies did annoy the sheep and cause them to seek the shade of cliffs, but usually flies were not much in evidence, probably because of the cool breezes on the ridges. Since insects are relatively scarce on purely summer and winter ranges, it seems unlikely that any difference in their incidence would be sufficient to cause migratory movements.

One factor which may be of importance in explaining migrations is the natural tendency of animals to wander. Let loose some horses and they will wander widely in their grazing. In any area the sheep wander about considerably over a period of days. Repeated wandering movements to certain localities may in time have established definite migration habits. In other words, the sheep may migrate simply because they like to travel.

The factors that originally caused the movements of sheep may have disappeared, and the animals may now be migratory largely because of habit which has no present-day use. When the population was excessively large, as it was in the late 1920's and no doubt has often been in the past, the sheep may have moved from an overgrazed winter range to the fresh pastures unavailable in winter. Now that there is no need for fresh pastures, the sheep may continue their treks because of a habit handed down to them.

With few exceptions the lambs are born on the winter

range. At lambing time, May and early June, the ewe seeks solitude in high, rugged terrain where she is safe from predators. Here she awaits the birth of her lamb and lingers with it for a few days before rejoining other ewes. Lambs two or three days old, so small that they can walk erect under their mothers, clamber up cliffs so precipitous that even the mothers can scarcely find footing. Not only can the lambs climb, but they possess unexpected endurance. I frequently saw very young lambs hurrying after their mothers from near the base of a mountain to its very top without resting.

The travels of a captive lamb illustrate the endurance they possess. On May 13, 1928, a ranger brought to his cabin a lamb only a few hours old. It soon became a pet and amused itself by leaping about on the chairs and beds. It followed the ranger and his cabin mate around all day and showed no desire to join the other sheep that were sometimes in view. It was raised with a bottle, on powdered milk. When only two and a half weeks old, it followed the men for thirty miles over rough ground and through glacial streams. When it was a month old, it became wet in a glacial stream when overheated and died of "pneumonia."

It surprised me when I first observed the speed of the lambs on relatively smooth slopes. On May 28, 1939, when most lambs were a little less than two weeks old, I came suddenly upon six or seven ewes and their lambs. My nearness gave them quite a fright, causing them to flee full speed down a gentle slope to the bottom of a draw and up the other side. Although the ewes galloped at full speed, the lambs kept up easily and at times two or three of them even forged out ahead. Upon reaching the bottom of the ravine and starting up the other side, the lambs seemed to have an advantage and "flowed" smoothly up the slope ahead of the ewes. When the group finally stopped to look around, the lambs were as fresh as ever and seemed anxious to make another run. This precocity of the lambs is, of course, of great value in avoiding wolves and other predators.

During the first week or two after lambing, the ewes are especially wary. When approached, they hurry off with their lambs, often crossing over to another ridge so as to leave a draw between them and the intruder, or else climb high into rough cliffs. Before the lambs are born, these same ewes might be rather tame, and as the lambs become older they again become tame.

The lamb is not left lying alone, as are the young deer, elk, and antelope. It remains near its mother, pressing so close to her at times when traveling that it may be almost under her. When resting, the young lamb usually lies against the mother or only a few feet away. This habit of remaining close to the mother is of high protective value should eagles attack.

I have observed eagles dive at ewes and lambs, but their maneuvers do not necessarily indicate the degree of eagle predation on lambs. Eagles have also been observed swooping low over grizzlies and wolves when there was no intent of predation. Once an eagle dived at an adult wolf which was standing near its den. About a dozen times the eagle swooped, barely avoiding the wolf which, each time, jumped into the air and snapped at it. The eagle turned upward at the right moment to avoid the leap and apparently was enjoying the game.

When they are a few weeks old, lambs move about with more freedom and gambol over the meadows in little groups. Judging from their behavior, they are not greatly worried about eagles after three or four weeks. In late June and early

July, I have seen eagles fly low over lambs separated from their mothers without attempting to strike and without alarming them. Throughout the summer the eagle may occasionally dip downward at the sheep just as it does at other animals. On September 10 an eagle swooped low over two ewes and a lamb, giving them quite a start. It sailed close over them several times, calling as it passed. But after the first start, the sheep seemed unafraid.

When the lambs of the Dall sheep are young, the mothers often congregate in rough cliffs. A few may remain with the lambs while the others go out three or four hundred yards to feed on the gentler slopes. On May 26, 1939, I saw thirteen lambs frisking about on some cliffs on a rocky dike on East Fork. With the lambs were six ewes — three lying down just above the group and three standing in their midst. While I watched, one of the ewes moved off 150 yards and commenced to feed. A lamb followed her for ten yards, then dashed back to the others. One lamb ventured out alone for twenty yards, and a ewe at once became alert and moved four or five steps in its direction, whereupon the lamb hastened back to the group. Four ewes were feeding on a slope about 250 yards away from the cliffs. Occasionally one of them would look toward the lamb assemblage. From a point about a quarter of a mile away, three ewes stopped feeding and came galloping across the face of a shale slope, calling loudly as they came. When they were still about a hundred yards away, two lambs galloped forth, each joining its mother and nursing at once. The third ewe did not have such an alert or hungry lamb, for she smelled of five lambs in the top group, then walked briskly down to the rest of the lambs fifteen yards below; there she found her lamb, which belatedly came forth to nurse.

On June 5, 1939, there were twenty-two lambs on this out-cropping of cliffs and an equal number of ewes, but some of the ewes were feeding two or three hundred yards away. One lamb on this occasion ran out about sixty yards to meet its mother and to nurse.

On July 12, 1939, a band of fifty-two ewes, lambs, and year-lings were feeding on some rather gentle slopes of Cathedral Mountain. The mothers frequently moved off one or two hundred yards from their lambs, which remained in a scat-tered group resting on the slide rock. Five or six ewes at different times stopped feeding and called loudly. Each time, a lamb would recognize the call of its mother and gallop down to nurse, usually for a minute or less. The sheep were quite noisy at times, the ewes "ba-a-ing" loudly and the lambs answering more softly.

The tendency of ewes with lambs to segregate is probably a natural outcome of their all having the same inclination to remain in the rougher terrain. Later, there is more inter-mingling of the ewes with lambs and those without lambs.

When a ewe leaves her lamb in a group and goes off to feed, she may lose her lamb temporarily. On June 7, 1940, I climbed to a small group of ewes and lambs, which ran out of sight when I neared them. I was standing on the spot where they had been when a ewe, which I had passed on the way up, approached me, calling. There was not much room on the cliff where I stood, and when the ewe was seven or eight paces from me, I stepped to one side to let her pass. She was evidently perturbed at finding me where her lamb had been, for she called loudly and came toward me with lowered head, which she jerked threateningly upward as though to hook me with her sharp horns. After I had warded her off with my tripod, she stood for a few moments calling, then hurried in the general direction the other had taken. No doubt she soon found her lamb, for the band had not gone far.

On June 13, 1941, I spent some time on Igloo Mountain, observing about fifty ewes and thirty lambs. At first, most of them were in a broad, grassy basin. One ewe nursed her lamb, walked off two hundred yards, and fed with another ewe for an hour, never once looking toward the lamb. As I climbed, a small group of ewes and lambs, including the lamb of one of the two feeding ewes, moved upward, feeding higher and higher until they were out of my view. When the two ewes rejoined the main band, which in the meantime had moved two or three hundred yards to some cliffs, one of them found her lamb at once but the other was not so fortunate. Calling continually, she searched through the entire band, duplicating her visits to some parts of it, but still with no success. She finally recrossed a gulch and climbed in the direction taken by the little group with which her lamb had gone. She showed good sense, and no doubt soon found her lamb.

After the sheep had rested in the cliffs for some time, another ewe suddenly commenced to call at short intervals from a prominent point. The ewe had called a long time when two lambs came forth from a large crevice in a rock a short distance away, where they had been lying. The ewe saw them emerge and hurried to join them. One of them was her lamb and it immediately nursed.

The lambs play a great deal, romping with speed and agility. Sometimes they butt each other, coming together after a two- or three-foot charge. Occasionally a ewe will play with the lambs. Once a ewe played with three lambs, dashing after one, then another. A lamb would leave its mother, get chased, and rush back again to its parent. Apparently it was great fun.

The summer season is a happy time for the sheep — the fresh, green growth is highly palatable, and there is little danger from wolves or other predators. In the fall, when the sheep have returned to their habitual winter ranges, they are in full vigor, fortified for a long winter, when food is less nutritious. Here the rams engage in spectacular fights in November and December — the mating period. It is chiefly on the winter range in the cold snowy months of the year that the sheep must be especially alert to guard against their enemies.

20. The Wolves Go Sheep Hunting

I SPENT MUCH TIME in the sheep hills in an effort to learn the hunting technique of wolves seeking sheep and the general behavioral relationship of these two species. I shall describe a number of incidents involving wolves and sheep, including some successful hunts. The hunting incidents are chiefly interpreted from tracks in the snow, which often showed clearly what the sequence of events had been.

On May 7, 1940, I saw a scattered band of sheep move slowly from one ridge to an adjoining one. The movement was so definite and consistent that I suspected that the sheep were moving away from danger. A little later I peered into the draw below the ridge first occupied by the sheep and saw a black wolf investigating some cleanly picked sheep bones. The sheep had simply preferred to have a ridge between themselves and the wolf.

While eating breakfast at the Igloo cabin on May 3, 1940, I heard a wolf howling nearby. Stepping outside, I noticed three alert rams, each on a pinnacle, peering intently below them. They continued to watch for some time, evidently keeping an eye on the wolf I had heard. They apparently felt safe where they were, even though the wolf was directly below them.

On August 3, 1940, Dr. Ira N. Gabrielson and I watched a

black wolf trot leisurely down a short draw on the ridge opposite us and descend the narrow stream bed bordered by steep slopes. Two rams on the slope below us watched the wolf, and when it trotted out of their sight, they moved to a point where they could again see it. Seven other rams grazing a short distance from the two paused but briefly to look. The wolf stopped a few times to look up at the rams, but continued on its way until finally hidden by a ridge. The rams and the wolf had shown a definite interest in one another but that was all. The wolf probably examines, at least cursorily, all sheep in the hope of discovering an opportunity for a successful hunt; and the sheep keep alert to the movements of the wolf so as not to be taken by surprise or at a disadvantage.

On June 29, 1941, about sixty ewes and lambs on the south side of Sable Mountain moved up the slope a hundred yards or more and stood with their attention centered on the terrain between us. A search with the field glasses revealed a gray wolf loping westward between me and the sheep, about a half mile from them. Some of the sheep began to feed, others watched until the wolf had passed. Because the day was dark and rainy, the wolf, whose legs and lower sides had become wet from the brush, was unusually hard to see, yet the sheep had quickly discovered it.

Former Ranger Lee Swisher told me that he had seen six wolves suddenly come close to seven rams feeding out on some flats at Stony Creek. The rams bunched up and the wolves stopped a hundred yards away. They made no move toward the rams, which, still bunched up, walked slowly and stiffly toward the cliffs. The rams maintained the slow gait until they had almost reached the cliffs, then they broke into a gallop and quickly ascended the rocky slope. The incident seems to indicate that a wolf may to some extent recognize the ability of the rams to defend themselves. The wolves on this occasion may not have been hungry; possibly, under other circumstances, they would have made some attempt to single out one of the rams.

On one occasion Mr. Swisher said he let a sled dog chase some rams. They turned and faced the dog with lowered horns. After thus threatening the dog, the rams started up the slope, and when the dog followed, they again turned on him. This incident is indicative of what the seven rams approached by the six wolves would have done had they been attacked.

A dog belonging to Joe Quigley, a miner, was said to have escaped one night from his camp in the sheep hills. The dog returned two or three days later, badly battered. Some time following this event, when the team was driven up to some sheep carcasses, this dog was not at all anxious to approach them. The inference was drawn that the bruises the dog had suffered during his absence had been administered by a sheep.

On June 19, 1939, I saw twenty-two ewes and lambs feeding among the cliffs a short distance above four resting wolves, one of which was lying only about two hundred yards away. The sheep had already become accustomed to the presence of the wolves when I saw them, for they grazed unconcernedly. Their confidence was probably due to the proximity of exceptionally rugged cliffs to which they could quickly retreat should the wolves attack.

On August 3, 1939, I saw a band of twenty sheep run up the slope of a ridge bordering East Fork River. A little later a wolf climbed the slope, making slow progress. Twelve sheep watched from a point up the ridge, three from some rocks not far from where the wolf went over the ridge top. Two eagles swooped at the wolf a number of times, continuing to do so after the wolf was out of my view, so that I could follow its progress by watching the eagles. The sheep quickly resumed grazing. They had not moved far from the wolf but had watched to see what it was up to.

On the morning of September 15, 1939, five wolves (the East Fork Band) trotted along the road to Igloo Mountain, then climbed halfway up the slope, which was covered with several inches of snow, and followed a contour level. I saw them a mile beyond Igloo cabin; three were traveling loosely together, a little ahead of the other two. Sometimes they were strung out, fifty yards or more apart. Generally they trotted, but occasionally they broke into a spirited gallop as though overflowing with excess energy. Opposite me they descended to a low point, and the two gray wolves which had brought up the rear dropped to the creek bottom. A black one rounded a point and came upon three rams, which it chased. The animals then went out of sight, but in a few minutes the wolf returned and I saw the rams descend another ridge and cross Igloo Creek. As they climbed a low ridge on Cathedral Mountain, they kept looking back. When crossing the creek bottom, they had not been far from the two gray wolves, which traveled a mile up the creek and then returned to join the others.

After chasing the three rams, the black wolf joined another black and the dark gray female, and all moved up the slope. One of them stopped to howl, possibly calling to the two grays on the creek bottom. These two turned back about that time and later joined the others.

On a pinnacle of a high ridge stood a ewe, peering down at the approaching wolves. She watched a long time but moved away while the hunters were still far off. The wolves went out of view on the other side of this same ridge. Later I saw some ewes farther along the ridge looking steadfastly to the west in the direction the wolves had last taken, and beyond these ewes three more were gazing intently in the same direction. The sheep had fled to the highest points and were definitely cautious and concerned because of the presence of the wolves, which seemed to be coursing over the hills, hoping to surprise a sheep at a disadvantage. The day before, I had seen a lamb in this vicinity with a front leg injured so severely the lamb was not using it. Such a cripple would not last long if these wolves found it. The habit of cruising far in his hunting gives the wolf opportunity to find weak sheep over a large range and to come upon undisturbed sheep, some of which he may find in a vulnerable location.

Charles Sheldon tells of following the trails of two wolves in March and finding that on nine occasions they had chased sheep unsuccessfully. On eight of the chases they had descended on the sheep from above. He said that the sheep in the region had become badly frightened and that "most of them kept very high." The occurrence of so many unsuccessful hunts suggests that these two wolves were testing out each band, hoping eventually to find a weakened animal or to gain some advantage. It appears that wolves chase many sheep unsuccessfully and that their persistence weeds out the weaker ones.

On October 7, 1939, I saw the track of a single wolf that had crossed Igloo Creek and then had moved up the slope of Igloo Mountain. After following the creek about a mile, I saw two ewes and two lambs on a spur of Igloo Mountain watching a black wolf which was curled up on a prominent knoll on the next ridge about two hundred yards to the west.

The ears of the wolf were turned in my direction, and to avoid alarming him I walked along as though I had not seen him until I was out of his view. Then I doubled back close to the bank and ascended the draw toward him. But when I came near the knoll he was gone, and the sheep were lying down.

After backtracking the wolf, I deduced that its actions were as follows. After crossing Igloo Creek, it climbed part way up the mountain. It followed a trail along a contour at the edge of some spruces, near the point where a wolf (perhaps the same one) had surprised a lamb among the spruces a few days before. The wolf crossed some draws and small spur ridges and arrived at a ridge on the other side of which the four sheep were feeding in a broad swale. The wolf climbed up the ridge, out of sight of the sheep, for fifty yards, so that he was slightly above them. He then advanced slowly until within 150 yards of the two ewes and two lambs and then galloped down the slope toward the sheep. The latter had escaped to the next ridge, from which they had been watching the wolf when I first saw them. The wolf chased up the slope after them but a short distance and then continued westward to the next ridge, where he curled up in the snow. This time the sheep had the advantage and escaped. It is significant that they did not run far beyond the wolf; apparently they were confident that they were safe, since there was much rugged country above them. The behavior of the sheep is definitely conditioned by the terrain they are in and their position in relation to the enemy. Approach a sheep from above and he feels insecure and hurries away. A sheep on a flat is much more wary and timid than one in rugged country.

On October 4, 1939, by backtracking a wolf, I found a

freshly killed male lamb. The tracks in the snow plainly told the story. The wolf was following a trail along the side of Igloo Mountain within the edge of the uppermost timber. Suddenly he came upon five or six sheep feeding among the trees. Those above the trail ran up the slope to safety, but a lamb which had been feeding farthest down found itself cut off from possible escape to the top of the mountain and was forced to run on a contour in the direction from which the wolf had just come. In chasing the lamb, the wolf was able to gallop back over the easy trail it had been traveling and thus keep above the lamb, which was endeavoring to swing upward in front of it.

The lamb traveled parallel to the trail over terrain broken up by the heads of numerous small draws. In one place he veered slightly upward until he came to the trail, but he must have been hard pressed, for he again turned downward and now the wolf followed him directly. The chase led gradually down the slope, the lamb apparently keeping as much altitude as possible in the hope of gaining the rocks above. But he was too hard pressed to cut upward ahead of the wolf.

Finally, the lamb descended a steep gravel bank to the creek bottom, crossed and recrossed the creek, started up the steep gravel bank at an angle, and returned to the creek, where he was killed. He had run slightly more than half a mile before being overtaken. This time the wolf had the advantage not only of coming suddenly upon the lamb from above, but also of having a trail to follow while the victim was galloping over rough, brushy country. If the sheep had been feeding in the open as they almost always do, the wolf probably would have been discovered before he was so close to them. This is an example of a situation where the predator gains an unexpected advantage.

On October 5, 1939, there were fresh tracks of foxes, wolves, and a grizzly in the snow, all leading up a small, cliff-bordered stream flowing into Igloo Creek. It was evident that there was some special attraction near at hand. My companion and I proceeded cautiously and soon saw a wolf run off. Farther on we saw a lone lamb in the rocks. An eagle flew away, and then we noticed a grizzly chewing on the skull of a ewe. I hoped the bear would leave the skull so I could get the age of the animal and examine the teeth, but when he finally moved off and climbed a low promontory, the skull was in his jaws. He soon finished crunching the bones, then climbed to a rock a little higher, where he lay down and after a few minutes went to sleep. The skull was a little close to the bear for us to retrieve, but that difficulty was solved by two magpies which, in fighting over it, knocked it off the cliff to a spot where we could safely get it. After considerable searching we found the skull of a second ewe on a grassy knoll a few feet above the gravel bed.

The two ewes had been killed fifty yards apart. Nothing now remained except a few large pieces of hide, some legs, entrails, stomach contents, and the two broken-up skulls. The assemblage of animals gathered at the kill may have consumed all the meat, or perhaps the wolves and foxes had cached what was not eaten.

The snow on the ground made it easy too backtrack the chase. The story was simple. The two ewes and a lamb had been feeding on *Equisetum* on a broad, moist swale not far above Igloo Creek. (The stomach contents of the two victims were made up mainly of *Equisetum*.) The wolves had been following along the mountain slope at a level slightly higher than that where the sheep fed. Coming over a rise, they spied the sheep feeding in the swale 150 yards below them. The tracks showed that the wolves did not start running until they were within seventy-five yards of the sheep. The latter galloped out of the swale and ran downward at an angle toward Igloo Creek, which they crossed after descending a steep dirt bank about a hundred feet high. The wolves followed directly after the sheep, but instead of running among the large rocks in the canyon stream bed as the sheep had done, they ran alongside the rocks on the smoother ground covered with a sod of avens. Two hundred fifty yards up this creek the two ewes were captured just before reaching some cliffs. The lamb escaped, and, as stated, we saw it alone a short distance above the carcasses. The ewes had run half a mile before the capture.

These two ewes had lived almost their full span of life, for one was ten years old and the other was eleven. The teeth had been used up and were no longer pushing out to compensate for wear on the surfaces. In one ewe the tooth surface was worn below the gum. In the other a molar had worn a "crease" in the palate. They were weak animals, less able to escape the wolves than a lamb. A lamb is itself in a vulnerable age class, apparently, but in the fall, before the hardships of winter have affected it, the lamb is probably almost as well able to escape a wolf as is an adult.

Late in the winter of 1940, in the rugged draws on the north side of the Outside Range between Savage and Sanctuary rivers, Harold Herning and Frank Glaser found much sheep hair, bloody pieces of hide, and wolf tracks, all indicating that the wolves had been killing sheep. No skull remains were found. The bones were, no doubt, hidden in the deep snow which filled the bottom of the draws. After the snow disappeared, Al Millotte (later an outstanding Disney photographer) and I hunted in these draws for skull remains and found the skulls of four recently killed sheep and the hair remains of a fifth. Two of the skulls were those of eleven-year-old rams, one of a twelve-year-old ewe. Here the wolves had apparently eliminated some animals doomed to die soon of old age.

On January 7, 1941, I saw the tracks of five or six wolves along the road toward Savage Canyon. A short distance above the canyon the wolves abruptly left the road and climbed a gentle slope to the top of an isolated rocky promontory. They chased a lone ram down its steep side to the creek bottom, where I found remnants of hide, as well as the stomach contents and the skull. Apparently the wolves had seen the ram on this isolated bluff and had turned aside to circle behind him and cut off his retreat to high ground. He was twelve years old, past his prime, a weak animal. The method employed in capturing him — that of coming down from above and driving him down the slope — seems to be a typical hunting technique.

Some observations made in the sheep hills bordering East Fork River on the morning of May 26, 1939, show that, on a relatively steep, smooth slope, sheep are easily able to avoid a single wolf. With a companion, I had climbed to the top of a ridge from which I had a view of some snow-free ridges on the other side of a small creek below me. I noticed a band of sheep resting on a smooth slope that slanted at an angle of about forty degrees or a little less. While I watched, the sheep bunched up and ran off about thirty yards to one side. Through the glasses I saw a gray wolf a short distance above. He loped toward them and the band split in two, some going upward around the wolf, the others circling below it. When the wolf stopped, so did the sheep, only thirty or forty yards from him. He galloped after the lower band, which ran downward and then circled, easily eluding him. Compared to the sheep, the wolf appeared awkward. After a few more sallies the wolf lay down, with feet stretched out in front. One band lay down about seventy yards above him, the other about fifty yards below him. Only one sheep in the lower band faced him; the others as usual faced in various directions. One sheep fed a little before lying down. The lower group consisted of five ewes, one yearling, and three rams. In the upper group were four ewes, four yearlings, one two-year-old, and two rams. All rested for one hour. Then the wolf again chased the lower band, which evaded him as before by running in a small circle around him. A flurry of snow then obscured my view. When it cleared a few minutes later, the wolf was disappearing in a draw and the sheep were grouped on the ridge above him.

In a short time he reappeared and slowly worked his way down the ridge to the creek bottom. Nine ewes, each with a lamb, appeared on the ridge near the draw which the wolf had just left. The lambs were at the time only a week or so old, but apparently they had been able to avoid the wolf. The utter lack of fear exhibited by these sheep is quite significant, indicating that a single wolf can easily be avoided on a slope.

The behavior of a ram chased by Charles Sheldon's dog is somewhat similar to the behavior of the sheep attacked by the lone wolf. Sheldon gives an excellent description of the event:

At that moment, a short distance ahead, I saw a three-year-old ram crossing the divide toward Intermediate Mountain. Here was a rare opportunity to observe the actions of a sheep when chased by a wolf. Quickly taking the pack off Silas, I led him ahead to within a hundred yards of the ram, which had not yet seen us.

Silas dashed at him full speed. The ram rushed toward the slope a hundred yards ahead. For a hundred feet the dog did not gain, but during the next hundred he gained at least twenty-five feet, and during the next hundred he was gaining rapidly. Although the dog was clearly the speedier of the two, I thought that the ram deliberately slackened his speed as he neared the slope, which was sharply inclined. The dog was not forty feet behind when the ram reached it. Up he went, bounding for forty feet; then turned and coolly stood a moment to watch the dog, which was running up at almost equal speed. Then the ram turned and rather leisurely ran upward a hundred feet, gaining somewhat on the dog, who by that time was going much more slowly. This time the ram stood and watched until the dog was within twenty feet, then easily ran up another hundred feet and again stood and looked at the dog. Silas, however, was now only trotting, and his panting showed that he could not run upward any more. Yet he followed the ram, which kept repeating the same tactics, never losing sight of the progress of the dog, until within a hundred feet of the crest, where a sharp projective rock rose almost perpendicularly from the slope. The ram quickly climbed to the top and looked down at the dog, which now was only walking. Nor did he move when Silas reached a point fifty feet below him. Then the two stood looking at each other. Finally, the dog turned and trotted back to us.

Not once, after the first burst of speed on the level, did the ram show any fright. When he knew he could reach the slope, he was deliberate in every movement, and after reaching it, he coolly played with the eager dog. After each advance, however, he was careful to turn and watch his pursuer. He seemed to know that the dog would soon give up the chase, yet I believe he did not credit Silas with the persistence he had displayed. The actions of the ram led me to suspect that a wolf would not have followed more than a few feet up such a slope, its experience, which Silas lacked, having taught it that a sheep could easily escape when once headed upward on a steep slope.

At several places, notably Igloo Creek, Polychrome Pass, Toklat, and a stretch opposite Mount Eielson, the automobile highway passes through winter sheep range. In some places it cuts into the heart of the more rugged cliffs utilized by the sheep. The highway favors the wolves in three ways. First, it gives them an easy trail along the entire winter range, so that they can move more readily from one part to another. Second, it gives the wolves easy access into the cliffs themselves; they need not make a laborious climb to get among the sheep but can follow a smooth, easy grade. Finally, each blind corner in the road — and there are many of them — is a hazard, for the sudden appearance of a wolf may give the sheep no time to escape. This is especially true when the sheep are bedded down on the road.

A ewe, lamb, and yearling were killed on the road at Mile Sixty-seven on September 20, 1939. The victims had been bedded down near a sharp corner. Four or five wolves had come around the corner, made a dash at the sheep, and captured them before they had run more than a few yards.

Several sheep killed by wolves were found on and beside the road at Polychrome Pass and at Igloo Creek.

While the road affects seriously only a small part of the winter range, it is a good illustration of disruption of natural wildlife relationships by an artificial intrusion.

In a high draw, three yearlings were found together which had been killed early in the spring. Two of them showed necrotic lesions on the mandibles and apparently were in poor health when captured by the wolves.

On the snow-covered mountain slope above our cabin at Igloo Creek on the morning of October 14, 1939, I saw a fox feeding, and on the same slope a raven and some magpies. Obviously there was a fresh kill. We climbed to the spot and found the remains of a nine-year-old ewe affected with severe necrosis of the jaws. The horns were short and stubby and perhaps indicative of a prolonged chronic ailment. The teeth were irregularly worn. Some were long and sharp, others were worn down to the gum. One tooth was shoved entirely out of line. A premolar was broken off. An upper premolar was bent outward. Cavities along the teeth were packed with vegetation. The animal had been alone on the cliff (sick animals are often solitary). Five wolves had killed the sheep on a steep slope below some cliffs. They had crushed the skull and eaten the brains. After feeding, the wolves had curled up in the snow on a rocky spur to one side of the remains, about a quarter of a mile from the cabin. There were seven beds, but only five of them were coated with ice. In the other two beds the wolves had not lain long. Although the ewe had been found in rough country under conditions favorable for escape, she was so weak that the wolves were able to make the capture.

A few generalizations can be made concerning the methods of the wolves in hunting Dall sheep. It is my impression that the wolves course over the hills in search of vulnerable animals. Many bands seem to be chased, given a trial, and if no advantage is gained or no weak animals discovered, the wolves travel on to chase other bands until an advantage can be seized. The sheep may be vulnerable because of poor physical condition due to old age, disease, or winter hardships. Sheep in their first year also seem to be specially susceptible to the rigors of winter. The animals may be vulnerable because of the situation in which they are surprised. If discovered out on the flats, the sheep may be overtaken before gaining safety in the cliffs. If there are weak animals in a band, their speed and endurance will be less than that of the strong and they will naturally be the first victims.

A wolf hunting alone can, apparently, be avoided easily by

healthy sheep on a slope. The lone wolf must find his animals at a decided disadvantage to be successful. Two or more wolves can hunt with much more efficiency. The method is to get above a sheep and force it to run down, for a sheep running upward can quickly outdistance the wolves and escape. Sheep on somewhat isolated bluffs, where space for maneuvering is limited, are in danger of having their upward retreat cut off and of being forced to run down to the bottom. I have found a number of carcasses in situations suggesting that the sheep had been chased down bluffs of limited extent, and my observations indicate that weak animals are the ones most likely to be found in such vulnerable situations. They often lack the energy to climb to more safe retreats to rest. Where the wolf population is relatively large, its pressure on the sheep probably is proportionately greater, eliminating a high percentage of weak animals and capturing more strong animals surprised at a disadvantage.

21. The Caribou Herds

THE CARIBOU is a circumpolar deer adapted to life in the Arctic. In this streamlined age, I suppose its rounded hoofs and long-muzzled, square-nosed face are black marks against it — technically. But come face to face with a fine old bull in his fresh autumn uniform! His great antlers sweep back and up, with one or two shovel-like brow tines over the muzzle carrying numerous lesser points. "The angularity of the antlers is a mark of beauty," you breathe, if you take time at all to form your thoughts. Your eye runs over the glistening white, massive neck, the low-hanging white mane, the whiteness spreading over the shoulder and trailing along the flank in a white stripe. You will catch the contrasting dark brown; it will appear almost black on legs and face. Yes, even a white nose, white anklets on all four feet above the black hoofs, a white tail patch, and white daubs on the tarsal glands. Suppose he hasn't seen you, and he wanders off slowly. Somehow it doesn't occur to you to call him odd. Instead, you may let your eye take in the reds and yellows of the ripening tundra, the sweep of the slope over there, a glimpse of the river beyond, the snow-capped mountains against the sky. This is the caribou at home. Contrary to the rule in deer, the females carry antlers — small and branched; and even the calves grow a spike six or seven inches long.

The caribou are gregarious; the herds sometimes number in the thousands.

The caribou are gregarious; the herds sometimes number in the thousands, but a few hundred is the more usual number. On the move much of the time, the animals make extensive migrations which may involve hundreds of miles of travel. The movements follow general route patterns over a period of years, but may have minor variations and sometimes rather drastic local changes. This migratory habit, along with the shifting of ranges, is highly beneficial to the vegetation, in that it tends to spread the use widely and thus lighten the grazing over the entire range. This situation is especially beneficial to lichens, the favorite caribou food, whose recovery is extremely slow when overgrazed.

There are five or more caribou populations in Alaska, each of which is sufficiently segregated to be considered a unit. The total population is now much less than in former years but still ample, and probably in general as large as is desirable from the standpoint of the lichen growth. The population in McKinley Park in 1959 numbered about nine thousand animals, an increase over the past ten years but a decrease from twenty years ago. They range over a region two or three hundred miles in diameter.

In recent years the park caribou have wintered chiefly north of the park, in the Lake Minchumina region. Each spring they move into the park in their trek to the Sable Pass area and beyond to the south side of the Alaska Range by way of the glaciers at the head of the Teklanika or Sanctuary rivers. When the herds arrive at Sable Pass in May or June, they cross broad snow fields, as though a relentless inner urge were driving them forward in their pilgrimage. As one watches the long lines methodically making their way, the animals appear dedicated and full of high purpose.

The herds that have crossed the Alaska Range, traveling in small groups no larger than one or two hundred, assemble on the south side for two or three weeks. Then, late in June or early July, they recross the range, now in large herds of a thousand or two. Once I counted five thousand in a single group. The great herds move westward along a series of low passes leading to the Muldrow Glacier and beyond. This spectacular scene may take place in two or three days or it may be more protracted, depending on circumstances.

At this time, if the sun is shining, one may find large numbers standing closely bunched on one of the broad gravel bars. The sun activates the beelike warble and nostril flies which, respectively, seek to lay eggs and hatch larvae on the caribou. The caribou may stand on the bar all day to avoid the flies, or they may assemble on a snowdrift or seek a high, exposed, wind-blown ridge where they can stand in the cool breeze. When a cloud floats across the sun on such days, the herd disperses hungrily to nearby green slopes to feed, sometimes stampeding as they break the huddle.

When these flies are abroad, one often sees a single animal standing tense, with its nose protected in a wet sedgy spot. At intervals, it will make a wild dash across the tundra to escape the flies that have found it, and then again stand tensely watching with head lowered.

When migrating, the caribou often travel in a broad front and leave behind them scores of parallel trails only a few feet apart. These trails are to be seen in many parts of the park. They are so deeply worn that, even if some are not used for many years, they remain clearly evident.

The calving of the caribou does not always take place at the same spot. Possibly the location of the principal calving grounds depends largely on where the caribou happen to be at the time the calving takes place. This is roughly from the middle of May to the middle of June. The fact that the same areas are used several years for calving may simply mean that the general movements of the herds are similar for those years. To give birth to their calves, cows generally wander off for varying distances from the main herds. At this time, a lone cow is almost sure to be near a newborn calf. Generally, these lone cows still carry one or both antlers. Within a few days, the cows with their calves join the main herds.

The calves are unusually precocious. On May 17, 1940, I found a calf in a spot which I had passed about three hours before, so that I knew it was not more than three hours old. There was a packed-down area in the snow about twelve feet across, and I noted a little blood in two places. The calf's legs were still moist. It managed with considerable effort to stand up, and after walking a few steps in the snow, it fell down. It tried walking several times, seeming to gain additional strength with each effort until, finally, it was able to follow me around, which it insisted on doing. It gave the typical guttural grunt. While I was with the calf, the mother circled about two hundred yards away.

The behavior of this calf was similar to that of a number of other newly born calves which I observed. In a day or two they can follow the mother, and I would guess that in about a week or ten days they are able to run almost as fast. On June 1, I saw calves, chased by a wolf, keeping up with the cows. Most of the calves were not more than two weeks old.

I found no indication of cows leaving their calves while going off to feed, as do antelope, deer, and moose. From the time the calf is born, and is able to follow readily, it usually remains beside or behind the mother as close as a shadow. This attachment is important from the standpoint of survival in a herd animal, for a great many calves would otherwise lose their mothers. I have watched thousands of calves, and my impression is that they do remarkably well in remaining close to their mothers. It is true that, when a large herd has been disturbed, a few calves become separated, but with several hundred animals milling around, it is surprising that more of them are not left astray. On several occasions I have seen cows and calves searching for one another.

On May 29, 1941, I observed an interesting incident in this connection. On the rolling tundra between Savage and Sanctuary rivers I saw a band of about seven hundred adult caribou with calves moving westward and then circling northward. They had moved away from two grizzly bears. After the caribou had moved a little less than a mile, I noticed that two calves were lost, a half mile from the main herds. One calf joined up with a cow and calf and followed in their wake for some distance, and then moved off in another direction, obviously searching for its mother, which in the meantime had returned from the main herd and soon found her calf. When the mother came, the youngster at once commenced to nurse vigorously. It had been lost for at least half an hour. The other calf wandered toward the herd, started for several cows, then circled back toward the place it had left. It then disappeared in a swale and I did not see it again. This calf had been searching for its mother and, after following the herd, had returned to where, apparently, it had last seen her. A cow, probably the mother of the second calf, was searching, but while I watched, she did not go back far enough to find the calf.

On July 1, 1940, a lone calf came running in my direction and stopped within fifteen yards of me. It was looking for its mother and came to investigate me. It later crossed some broad gravel bars and found her.

The cows are reasonably solicitous when their young are in apparent or real danger. On May 24, 1939, I approached a cow with a calf that was less than a day old. When the cow ran off, the calf followed slowly. Part of the time, the mother trotted slowly enough for the calf to keep up, even though I was hurrying after them about seventy-five yards behind. Each time the mother found herself a short distance ahead, as happened four or five times, she returned and nuzzled the calf. When it finally lay down, the cow stopped about fifty yards away and did not run off until I captured the calf. Then she ran over a nearby rise and circled above me, keeping watch from a distance of about half a mile until I left.

This mother seemed more solicitous than a mother elk would be.

On May 24 two rangers and I chased a calf which was following a cow slowly. When one of the rangers overtook the youngster, the mother, a short distance ahead, turned and showed her defiance by pawing the air. After they ran off, we caught up to them and another cow and calf, and both calves dropped in the brush when they were unable to keep going. The mothers peered anxiously at us from the steep slope opposite. After going over the hill, one of the cows returned and came to within seventy-five yards as we were departing. Some cows run off more readily than others, disappearing after a brief show of solicitude. Quite often they return to peer over a hilltop to see if the intruder has gone.

On occasions, when wolves have killed a calf, I have seen the mother searching the area for it. Once I observed a cow smelling of her calf at least several hours after it had been killed. On one occasion a cow with a calf too young to follow ran off with the herd but returned to the calf in ten minutes.

During the calving period the wolf, grizzly, and smaller predators prey to varying degrees on the calves. The mother must flee from the wolf and grizzly but perhaps can fight off some of the smaller animals. That the caribou is not entirely defenseless was well illustrated by a young bull kept in captivity by my brother and me in 1923. He showed much dexterity in the use of his hoofs. When we tried to force him to one side of a corral with a rope, he struck the rope on the ground fiercely and accurately with fore hoofs and then with hind hoofs. This bull was full of spirit and often chased us when we came near. Only the heavy toggle to which he was tied stopped his rushes. This exhibition in the use of the

learned that a necrosis had severed one jaw bone and formed an exostosis on the other jaw bone. Disease had doomed this calf. Lone calves probably do sometimes fall prey to eagles, but it seems likely that a healthy calf with its mother is seldom attacked.

The grizzlies that happen to be living where caribou are calving have opportunities to capture calves too young to escape. After the first few days, a calf is probably safe from grizzly attack, for it rapidly gains strength and speed. The total effect of grizzly predation is not large, for there seems to be no special movement of grizzlies into a concentrated calving area, and the few bears present would not take a heavy toll. Some of the calves eaten would no doubt be stillbirths or victims of other predators.

22. Wolves Hunt Caribou

Another enemy of the caribou is, of course, the wolf. Despite the fact that there is much wolf predation on adult caribou, one must be cautious and examine all accounts of such predation critically. A surprising number of reports, both written and spoken, do not bear scrutiny. One evening in 1939 some boys told us that they had just witnessed two wolves pulling down a cow caribou. They gave a vivid account of the incident and told us where the hunt had occurred. After supper three of us set out to examine the carcass. We found it, but the animal had been dead at least a week. The hair slipped all over the back and even on the legs. The boys had probably seen one or two wolves at the carcass and had made a good story of it.

A similar incident concerns a man who had come into Fairbanks from the direction of Circle with a description of how a pack of wolves were killing great numbers of caribou. The story passed many lips. A man interested in wolves hunted up the observer to get particulars. The observer said he saw the tracks of a band of wolves and one caribou they had killed. For the sake of accuracy, he was asked if one wolf could not make many tracks. The observer admitted that one wolf might make as many tracks as he had seen. He was asked how he knew the wolves had killed the caribou and if it wasn't possible that hunters had killed it, since it was near the road. The observer did not know about that. In this instance, although the wolves may well have been killing many caribou, the observer had little evidence on which to base his assertions. It illustrates how an erroneous evaluation of a situation from the quantitative standpoint may be built up in the public mind.

It is well known that wolves kill adult caribou, but it is difficult to learn what proportion of the caribou killed are below standard in strength. It is hard to know how nip and tuck the relationships are between the two species; how many healthy caribou chased by wolves escape, and how many succumb. One observer found that the wolves in the Brooks Range appeared to be getting chiefly the ailing adults.

In the spring the wolves prey extensively on the calves. The first day or two after birth the calves cannot run fast enough to give the wolves a chase, but in a few days they can almost keep up with the cows and then they force the wolves to do their best. In no instance did I see the wolf stalking caribou. Such maneuvers are unnecessary, for the wolf has no difficulty in approaching to within a few hundred yards of them. Generally, the caribou seem not to be worried much by wolves unless chased. I frequently noted caribou bands watching the wolves when they could have been moving away to a more secure position.

The wolf's method of hunting calves seems to give an opportunity for the elimination of the weaker animals. Usually the wolf chases a band of cows containing several calves. Because the speed of the calves is only slightly less than that of the wolf, at least on level terrain, they make the wolf do his best, and the chase continues long enough for a test of the calves. The weakest, the one with the least endurance, falters after a time and drops behind the others, and this is the one the wolf captures. In some instances, I suppose, a calf falls behind because it is younger than the others, but after these animals are a few weeks old, the differences in time of birth probably are unimportant, and the one actually weaker than the others is the one that succumbs. Thus the wolf appears to be a factor in maintaining quality in the herds.

There may be more weak animals in populations than has been generally realized. In this connection some observations of the elk in Jackson Hole, Wyoming, are pertinent. In the spring of the year there is much variation in the strength of the yearlings. If a large herd is started running, some of the yearlings may be seen to collapse, exhausted by their efforts. Most of these animals at this time are able to rise again after a rest, being weak only from the winter hardships. Others are apparently diseased, for they die soon after tumbling. In the presence of a predator, probably these weaker animals would be eliminated as the winter progressed. Likewise, in the Yellowstone I learned that many of the deer fawns died during the winter from general weakness or disease.

On June 16, 1939, I had my first full view of a wolf capturing a caribou calf. My assistant and I were sitting on a ridge high enough above the river to give us a good view of the prospect before us. We were classifying, according to sex and age, the bands of caribou passing up the river. All day, band after band passed us, going up the west fork of the Teklanika River to the glaciers of the high Alaska Range; others were coming down this fork and going up the east fork.

From where we watched, we could see bands of Dall sheep scattered among cliffs across the intimate, narrow valley, some resting, some feeding on green ledges. During a lull in our counting we spied a fox along the edge of a small patch of trees, and presently we watched several fox pups as they played nearby. Once a grizzly emerged from some trees on the flat and, for some time, grazed on vegetation, then disappeared while our attention was directed elsewhere. We saw a second grizzly on the slope across the river from us, but soon he also wandered out of view. The knoll on which we sat was decked with flowers, chiefly the yellow-centered white flowers of the mountain avens. It was an idyllic day in remote surroundings.

About noon, drama entered that was to make this day memorable. In the distance we noted a herd of about 250 caribou, chiefly cows and calves, coming downstream. Soon they were near enough for us to make out that they were galloping. Suspecting that they were being disturbed, I looked through the field glasses and saw a black wolf galloping after them. When the caribou reached the triangular flat between the forks of the river in front of us, the wolf was close upon their heels. The caribou in the rear fanned out so that they were deployed on three sides of the wolf. He continued straight ahead, continuously causing those in front of him to fan out to either side, making an open lane through the herd. Those on the sides stopped and watched the wolf go past. Soon most of the caribou were on either side of the wolf's course.

On the flat the wolf stopped for a moment, and so did all the caribou. Then he continued straight ahead after a band of about thirty; and these again fanned out, whereupon he swerved to his left after fifteen of them, which then started back in the direction from which they had come. The wolf chased these for about fifty yards and then stopped. Small bands of caribou, some of them only a hundred yards away, almost surrounded him. It seemed strange that they did not run from the vicinity of danger. Then the wolf seemed to come to a decision, for he started after twenty-five cows and calves farther from him than those he had been chasing. Before they got under way he gained rapidly, but soon they were fleeing.

For a time the race seemed to be going quite evenly, and I felt sure the band would outdistance their enemy; but I was mistaken. The gap commenced to close, at first almost imperceptibly. The wolf was stretched out, long and sinewy, doing his best. Then I noticed a calf dropping behind the fleeing band. It could not keep the pace. The space between the calf and the band increased, while that between the calf and the wolf decreased. The calf began to lose ground more rapidly. The wolf seemed to increase his speed a notch and rapidly gained on the calf. When about ten yards ahead of the wolf, the calf began to veer from one side to the other to dodge him. Quickly, the wolf closed in and at the moment of contact the calf went down. I could not be sure where the wolf seized it, but it appeared to be about at the shoulder. The chase had covered about five hundred yards and the victim was about fifty yards behind the herd when overtaken.

In a few minutes, the black wolf trotted a short distance to meet a silvery-maned gray wolf which was limping badly on a front foot. Together they returned to the dead calf, sniffed it, then moved off and circled to the left side of the ridge at the forks and climbed it slowly. Halfway up the slope they rested for half an hour, and then they continued to the top of the promontory, about a thousand feet above the river bars. On the way they flushed an eagle, which circled and twice swooped low over them. The wolves lay down on the point of the ridge, where they were still resting at seven P.M.

At this time we decided to examine the calf carcass and descended from our ridge, taking advantage of a high bank along the river to keep out of sight of the wolves.

When we reached a point on the bar opposite the kill, we saw the limping gray wolf coming down the slope. Then we saw the black one feeding on its prey about three hundred yards up the slope from where the kill had been made. He had no doubt carried the carcass while we were walking along the bar. He made several runs at six or seven magpies which were feeding with him. Soon the wolf left, returning once to chase the magpies from the meat. The gray animal, which had been lying a short distance above the black one, then approached the meat and carried off a large piece. When we reached the spot, there were only a few ribs and some entrails left. From a distance, as we walked toward camp, we again made out the grey female resting on the strategic lookout, on the point of the ridge between the two forks of the Teklanika River. She no doubt saw us too.

The following day we returned to the forks of the Teklanika in hopes of again seeing the wolves. A band of fifteen caribou, without calves, trotted along the bar, followed by a trotting gray wolf. When the wolf stopped to sniff at a fox den, the caribou stood watching him from a distance of only seventy-five yards. The caribou moved off and the wolf disappeared, perhaps to rest.

Later in the morning (about ten o'clock) we saw eight or nine cows and four calves galloping across the river bar, followed by a gray wolf loping easily. They all crossed the east fork of the Teklanika River and came out on the flat where, the day before, a calf had been killed. The wolf galloped rapidly across the flat after the fleeing caribou, which, with a long lead, reached the rough country at the base of the ridge.

The wolf gained on the caribou while they ran up and down the slopes and it ran on the level, but when the wolf also reached the rough country, it was quickly left behind. On top of the first slope, it gave up the chase after running somewhat over half a mile.

On June 1, 1940, from a lookout near the East Fork wolf den, I saw an interesting hunt. At this time there were many bands of cows with calves, some feeding only a quarter of a mile from the den. One group walked within fifty yards of it. From the lookout I counted 1,500 caribou. At four-forty P.M. the black-mantled male looked into the den and then walked down to the bar. He was followed closely by Grandpa, and shortly the black male came out of the den and also followed. It appeared that the black-mantled male had looked into the den to let the black male know he was going hunting. The three disappeared in a ravine leading up the long slope. Grandpa was limping badly on a hind leg, not using it at all when he galloped. The gray female started late, after the others were in the ravine. The black female was left to watch the den, resting near the entrance.

Far up the ravine the black-mantled male, followed closely by the black male, appeared on a large snowdrift. The black-mantled male waited for the black male and, when it came up, jumped and romped with him. The wolves seem to enjoy romping on these late spring snowdrifts, and I have seen mountain sheep, caribou calves, and grizzly cubs also jump about and play on them. The female wolf was following a couple of hundred yards behind. The black-mantled male turned southward at right angles and followed a bench. The others turned also, but about one hundred yards lower down the slope. Far in the rear appeared Grandpa, still limping badly. He turned about a third of a mile below the others, being out of their sight on the slope below the bench on which they were traveling. They moved southward, the black-mantled male loping in a rocking-horse fashion, apparently from excessive spirits. Several bands of caribou in front of the advancing wolves galloped rapidly up the steep slope. A mile or more to the south I lost sight of all except Grandpa, who stopped and howled.

Soon the others had swung around Grandpa, who acted as a pivot, and they all moved northward again toward the den. The black-mantled male was just below Grandpa, and far down the slope came the black male and the gray female. By the time these three came abreast of Grandpa, about two hundred caribou in one band, and some smaller bands, were galloping northward ahead of them. Some of the caribou ran up the slope. The larger band was followed by a grizzly, galloping below and parallel with it. The bear seemed to be hurrying to get away from the general commotion. He veered off to the river bar and there stood up on his hind legs and looked up the slope in the direction from which he had come. Then he dropped to all fours and continued across the bar.

The wolves stopped soon after they had started northward, and the black male howled. When the black-mantled male answered him from up the slope, all the wolves assembled on the high point where he stood. They lay down for a few minutes, then the black male moved down the slope at an angle and chased some caribou. For a time the caribou did not run, and the black male was well within 250 yards before they began to flee. He galloped hard up a low ridge and down into a shallow ravine, where he captured a calf after following it in a small half circle. In about ten minutes the black male came out in the open and howled, whereupon the three wolves on the point started toward him. The black male trotted to-

ward the den, turning aside on the way to follow the fresh
tracks of a cow and calf for several yards. The black male arrived at the den at six-twenty P.M., one hour and forty minutes after leaving it. He disappeared in the den for a few
minutes. Soon the other three wolves returned and the gray
female immediately went into the den, while the two gray
males walked up above to lie down. The last three wolves ar-
riving had not gone to the kill but had come directly to the
den.

The first part of the hunt seemed to follow a system of
herding the caribou, but after the wolves assembled on the
point, three of them took no further part in the hunt.

The next hunt that I watched was preceded by much tail
wagging and howling. After the ceremonies, at seven-thirty
P.M., the three males trotted across the bar westward parallel
to the highway. The two females remained behind, lying a
few feet apart near the den and watching the departing hunters. The males were out for the regular night hunt. All day,
caribou had been in the vicinity of the den, but the resting
wolves did not molest them.

Soon after their departure, caribou on the flats ran off in
various directions, showing that they had seen the hunters.
The wolves kept trotting southwestward. The black male,
ahead and to the right, soon passed out of my sight behind a
ridge. The black-mantled male was far out on the bar, and
Grandpa was out of sight near the black male. The wolves
crossed the neck of tundra between two forks of the East
Fork River. In the meantime, I hurriedly returned to the
road and drove westward in my car, stopping on Polychrome
Pass high above the rolling tundra over which the wolves
were traveling.

Small bands of caribou were scattered over the tundra below me. Now the black male was far ahead of his two companions. As usual, he seemed to be doing most of the hunting. He approached two or three bands in his course and watched them while they ran away. In these bands there happened to be no calves, and I wondered if the wolf was looking over each band to see if calves were present. The two grays caught up with the black male, and part of the time the large black-mantled one was in the lead, trotting gaily and briskly with tail waving. Once he dashed at a band, then stopped to watch. The scattered caribou came together in a bunch and ran off. There were no calves. Once the black male galloped hard after a herd but stopped to watch when he was near it.

As the wolves continued traveling, the mantled male lingered far behind and from a knoll in the tundra raised his muzzle and howled. He was answered by the deep, hoarse howl of one of the wolves in the lead. After traveling five miles, they again were together and, as yet, had made no serious effort to kill caribou. There were many calves in the country, but the small bands containing fifteen or twenty caribou which the wolves had encountered along the way happened not to have any. It appeared that the wolves were searching for calves. At ten P.M. the wolves crossed the road and went out of sight behind a ridge. I had seen only preliminaries of their night hunting.

Some of the bands that ran from the wolves went off only a few hundred yards to one side. Others, which went straight ahead in the course taken by the wolves, ran as much as a mile. Some bands fled because they saw others run, and on a few occasions took a course nearer the wolves.

On June 5, 1940, at five forty-five P.M., I saw the black-mantled male as I was going toward the den lookout. He was lying on a knoll, howling, a mile from the den. To avoid dis-

turbing the wolves, I retraced my steps. For five minutes after
I reached the road the wolf continued howling at short intervals, and then he trotted briskly toward Sable Mountain.
When I arrived at a point where I could get a view of the
slope, I saw about 250 caribou, including calves, running hard.
Then I noticed the wolf feeding on something, probably a
calf caribou. Whether he had just made a kill or had returned to an old kill I do not know, but I suspect that it was
an old kill. He fed about ten minutes, then lay down beside
the carcass and stretched out on his side. At intervals of ten
or fifteen minutes he raised his head for a look around.

While the wolf was lying there, small bands of caribou
passed near him. One band of ten adults and three calves
passed within fifty yards of him. After these had passed, he
looked up, then galloped easily after them for a hundred
yards, stopped, and after watching them a few moments
slowly returned to his resting spot. The cows and calves fled
full speed along the base of the mountain, a calf leading the
flight. At nine o'clock the wolf trotted slowly westward.

In the dim twilight of June 13, 1940, south of Polychrome
Pass, two wolves, which appeared to be the black-mantled
male and Grandpa, were harassing some bears which probably had raided their kill. After half an hour the two wolves
trotted westward and from above approached a herd of about
two hundred caribou, which included many calves. Both
wolves galloped hard toward the caribou, which angled up a
slope. After a few hundred yards of running, the rear wolf
stopped, and then chased a lone adult caribou which was
standing nearby. The wolf chased this caribou for two hundred yards, and then the wolf started up the long slope toward the herd its companion was chasing. Near the main
Alaska Range, where the chase led, the shadows were so deep

I could not see all that happened. But after the herd had gone almost up to the rugged slopes, it had turned westward to a broad flat between two ridges. A little later the wolves were chasing the herd up this relatively level flat to a point near the head of the valley, but then the animals were again lost in the deep shadows and I could not see if a kill was made. A half hour later the two wolves were back harassing the bears. The early return of the wolves suggested that they had not been successful, but in any event the caribou had given the wolves a long chase.

Usually the caribou did not run far from a wolf unless pursued, but on June 16, 1940, I saw a cow and calf galloping across the east side of Sable Pass. Evidently they had winded four wolves which were approaching. The caribou traveled a mile and a half or more while we watched and, still hurrying, disappeared behind some hills. Later the four wolves appeared, but they were not on the trail of the cow and calf.

To hunt, the wolves had gone to Teklanika Forks, ten miles or more from the den, yet, during the day, at least four hundred caribou had been feeding a mile or two west of the wolves at the den and at least two bands with calves had passed within a quarter of a mile of them. Wolves seem to enjoy traveling and perhaps have favorite hunting grounds which they seek.

On June 17, 1940, I saw only the two females at the den. Grandpa was seen at Teklanika Forks ten miles away. The black female seemed restless all day; possibly she was hungry. At five-thirty P.M., as it commenced to rain, she trotted to the den, then over to the gray female lying fifty yards away, and, after stopping with her a moment, trotted across the river bar. In about five minutes she appeared south of me, about a mile from the den, chasing a large band of caribou containing many calves. Some of the caribou ran off to one side and soon began to feed. A calf brought up the rear of a group she was chasing. When it appeared that the wolf might overtake the calf, most of the band and the rear calf veered upward to the left and seemed to increase their speed. The wolf singled out another calf, which was running straight ahead with four or five adults, but in a moment the chase went over the ridge. Then it commenced to rain so hard that the visibility became too low to see. It seemed to me that wolf might be successful.

On June 19, 1940, I observed the five adult wolves near the den from eight-thirty A.M. to six-ten P.M. A half mile north of the den at eleven A.M. one of the black wolves chased a band of thirty-five cows and calves for about four hundred yards and then gave up without catching any calves. During the day five bands of cows with calves, averaging about a hundred animals in a band, passed within a third of a mile of the den without being molested.

On June 22, 1940, at eight thirty-five P.M., the black-mantled male and the black male followed the river bar southward from the den. About three miles away two or three hundred caribou were feeding on a grass-covered flat. For about a mile the two wolves trotted together; then the gray one fell far behind. He moved along the east bank while the black one trotted briskly across the river bar diagonally toward the caribou. When he was about two hundred yards from them, he watched for about a minute as though to size up the situation, then started galloping forward. He ran in such a way as to drive all the caribou off the grass-covered flat toward the gravel bar. He did not try to catch any of them but was definitely herding the scattered animals. When he had run the length of the scattered herd and had all the caribou galloping out on the bar, he swung around the front end of the herd and then came back, chasing them all before him. As the wolf caught up with a group of caribou, they would veer off to one side. Then he would continue straight ahead to the next little band, which in turn would veer off to one side. Finally, he stopped, sauntered over to the bank, wandered around as though investigating the area, then trotted across the bar in the general direction of the den. He went into some willows, where he probably lay down, for I saw him no more. The caribou moved on westward along their migration route, feeding as they went, behaving as though they had completely forgotten the chase.

The black-mantled male wolf in this hunt lingered on the other side of the bar. Some of the caribou which had been driven out on the bar had drifted over near him, and he had chased a band up in the tundra. They ran far to the east, but the shadows were so deep that I could not follow the hunt closely enough to learn whether any calves were killed. I was able to see the running caribou, but caught only an occasional fleeting glimpse of the wolf.

I do not know whether the hunt had followed a general pattern of cooperative maneuvering, but it might be so interpreted. In this case there was no great advantage gained by the wolves, but under different conditions the maneuvers could be advantageous to them. If the black wolf had chased the caribou toward the gray wolf far enough to tire the caribou somewhat, the gray wolf could have taken up the chase fresh.

On June 23, 1940, at nine-thirty A.M. I saw 250 or more cows and calves running hard, a mile or more east of Toklat River. A wolf, apparently the black male of the band, was chasing them. The wolf chased one group after another until he finally had the various groups running in different directions. Although galloping hard, he did not bear down on any herd. It looked as though he was testing the groups, looking for a specially vulnerable calf. After considerable chasing, the wolf ran after four adults and one calf, driving them off by themselves. The calf broke off to one side and kept veering as though trying to return to the herd; in so doing it lost ground, for the wolf could then cut corners. When the wolf was about twenty yards behind the calf, it was unable to reduce the gap for some time, but the calf began to zigzag and lost ground. The wolf gradually reduced the distance to a few yards, but still the chase continued for another two hundred yards or so. For a time I thought the calf might escape, so well was it holding up. But the wolf finally closed in, and the calf went down. While the wolf stood over the calf, apparently biting it, it jumped up suddenly and ran for seventy-five yards before it was again overtaken. A few minutes after disposing of the calf the wolf trotted a short distance toward the herd, then returned to his prey. The caribou herd continued on its way westward.

On June 25, 1940, at about five P.M. more than two hundred cows and calves came out on the bars above the East Fork wolf den. They were strung out in a long straggly line, feeding as they moved. A wolf howled from a short distance above the caribou, and soon its howls were answered by two or three wolves at the den. Although the caribou were between the lone wolf and those at the den, they continued feeding. I could not see that any of them heeded the howling. About this time a heavy rain obliterated the view; it continued raining all evening, and I could make no further observations.

When the calves are only a few days old, the wolves can kill them with little effort. On May 29, 1941, I found two dead calves, probably two or three days old. They were twenty-five yards apart, between Sanctuary and Savage rivers, where hundreds of calves were being born. Birds had fed a little on each. Bloodshot wounds on neck and back made it plain that the calves had been killed by wolves. Neck vertebrae of one were crushed. Very likely the two calves had been killed about the same time.

Less than half a mile from these two carcasses I saw a lone cow smelling of a calf. She walked away a few steps and returned to smell again. Then she moved off two hundred yards to feed. I walked to the calf and found that it was dead. On skinning it, I found tooth marks around the head and back, which apparently had been made by a wolf. The calf was only a day or two old. The three calves had been killed within the preceding twenty-four hours.

On June 29, 1941, at about three P.M., my attention was attracted by a band of about four hundred caribou running over the rolling tundra a mile west of the wolf den. The black male wolf first ran toward one end of the band so as to chase the caribou forward. The herd broke up into groups of fifty or sixty, the wolf dashing along in the middle, and I could not be sure of the status of the chase. Then the wolf started after fifty cows and calves. There was a chase of about half a mile, and the wolf kept closing in upon the herd. Once he stumbled as he galloped, and rolled completely over. But he was quickly on his feet, and little time was lost. Then a calf dropped behind the others. This seemed to encourage the wolf to put on added speed, and in less than a quarter of a mile he overtook the calf, knocking it over as he closed in. The wolf was hungry and fed for about half an hour. This calf was captured more readily than usual.

Foreman Brown of the Alaska Road Commission camp told

me that on June 29, 1941, at Stony Creek, he had seen a gray wolf with a crippled hind leg chasing calves with no success. After a while the wolf moved off and waited for the herds to approach. But while the caribou were still some distance away, it jumped up and gave chase. It had good speed for a short distance but quickly tired and fell behind. It caught no calves while Mr. Brown was watching.

About ten A.M., July 19, I saw the black female wolf on the East Fork bar circling back and forth with her nose to the gravel. She made a sweep of a hundred yards downstream, then returned upstream and finally waded the river. After she crossed the stream her ears were cocked forward and she started on an easy lope up the creek, apparently focusing her attention on a definite point. I looked ahead of her and saw a caribou calf lying on the bar beside the creek, watching the wolf. When the wolf was about 150 yards away the calf jumped up and galloped upstream. It crossed and recrossed the creek a dozen times, and every time the wolf followed. The calf did not seem to be as speedy as much younger ones had been, for the wolf was running rather easily and gaining. A full half mile from the start the calf, now hard pressed, wallowed in a deep part of the stream and on the next leap stumbled and fell. In a jump or two the wolf caught up and pulled the struggling calf, which once gained its feet for a moment, across the deep part of the stream to the shore, where it was quickly killed. The main herds having passed westward some time before, this calf was a straggler and may therefore have been a weakling. Upon examining the gravel where the wolf had been sniffing around in circles, I found fresh calf tracks. I believe the wolf had not seen the calf before I arrived, because a short while before she had been observed three miles up the road.

23. Cranes and Caribou

THE ARTIST had dipped his biggest brush into a huge bucket of autumn paint marked "Shades of Red," and the intense green of summer was magically transformed to brilliant red, with touches of bright yellow on the cottonwoods and the many species of willow. Only the patch of dark spruce woods south of Wonder Lake and the pioneer lone trees in the adjoining open tundra lent reality to this vivid north country. From where I stood, I could look across the numerous intersecting channels of the McKinley River upward to the top of Mount McKinley — Denali, the most high — silhouetted against a blue sky. In such a riotous landscape could anything but color be experienced? Was not everything on so gaudy and grand a scale that all else in nature would seem insignificant?

Then I heard a familiar cry — it was the sky music of sandhill cranes in migration. I scanned the upper atmosphere hurriedly, for I wanted to watch the cranes as long as possible before they should pass by and disappear into the distant blue to the south. So high do they sometimes fly that it often requires some looking into higher altitudes to discover them. More clarion notes, now more distinct as the flock came nearer, and then I discovered the V-shaped lines of the cranes following along the crest of the Alaska Range. To the

Then I heard a familiar cry — it was the sky music of cranes in migration.

major V formation of the flock were joined subsidiary lines forming subsidiary V's. Behind these were small, independent V patterns, all integral parts of the flock, numbering about four hundred. And as these birds moved majestically eastward past the top of Denali, some 20,300 feet in elevation, I discovered other large skeins in their wake, westward as far as the eye would take me. Flock after flock was flying along the mountains, some just behind the peaks of Denali, some over the top, and still others on the near side, silhouetted against the snow-white mountain, the highest in North America. The geometric formations of the flocks varied greatly. Some were more loosely organized — a number of birds flying in groups of two to five, and instead of a V only a long line. One could easily imagine that this was one of the most thrilling days for the cranes — a gala northern pageant.

When the first contingent arrived over Muldrow Glacier, the 35-mile river of ice coming eastward off Denali, the formation broke as the birds began to circle. The calling became intense and exciting. The cranes had entered an upward air current, and as they circled they were lifted up. After the birds reached the top of the updraft, or perhaps for some other reason, the flock began to re-form, V formations developed, adjustments in position were made, most of the clamor of notes subsided, and, in orderly formations, with a few slight adjustments continuing, the flock was again winging on its way.

The second flock broke formation in the same updraft, and thus each flock took advantage of the rising air to increase its elevation. Perhaps this increased elevation allows the cranes more leeway for sailing, giving them a downhill ride. One assemblage flying low, only a few hundred feet above the river bars, also broke formation over the glacier and in its circling was carried from an elevation of about three thousand feet to twenty thousand feet in a matter of minutes.

Apparently the recent rainy weather had grounded many flocks, which today were all well fed and poised for southward flight in the clear weather. Cranes which had been widely spaced in their nesting activities, some in western Alaska, others in Siberia, were in more simultaneous flight than I had ever witnessed. As a biologist I should have meticulously counted the cranes, but prosaic counting today seemed a sacrilegious thing. I did count quickly the cranes in a few flocks and estimated the number of flocks, but only roughly, and I guessed that five or six thousand had taken part in this adventure. In the afternoon, when I was observing caribou, about two thousand more flew high over Wonder Lake. One flock of about six hundred moved in wide, wavering lines, perhaps too many for orderly V formations; or perhaps, since the last updraft, they had not had time to get aligned.

Words are inadequate to describe the flight, the many variations in the formations, the alternate beating of wings and sailing, the beauty of the flocks in silhouettes against the white mountain and the blue sky, and the exhilarating poetry of it all in this primeval wilderness country.

On another day in this autumn landscape I had gone to Wonder Lake to photograph caribou bulls, now in the fresh and striking attire of the rut. Each fall a few bulls may be found at Wonder Lake, waxing on green forage and lichens and waiting for the inner stir of the rutting period. Two old bulls, accustomed to our presence, were in preliminary stages of the rut. They had cleaned the velvet off their antlers by vigorously rubbing and thrashing spruce saplings and willow brush. The white antlers, at first tinted pink with blood from the velvet, were now stained in shades of rich brown acquired from the dye of bruised bark. Occasionally the two bulls clashed their tall, angular antlers in light sparring, but then they would separate and go to grazing. Once they stopped to chew on an old disintegrating antler. The shed antler, an interesting item in the tundra country, is also, because of the minerals it contains, a part of caribou ecology.

While my companions and I were with the two bulls, a herd of caribou appeared over a low rise at a trot. They saw the two bulls and stopped to look. The larger of the two walked toward the herd, and from the herd emerged its largest bull to meet the oncoming antagonist. This large bull was ready to do battle, and without preliminaries the two champions clashed. A sharp prong must have jabbed the herd bull, for he jumped aside, stood briefly, and then moved away, followed by his loyal herd, and all were quickly out of sight in the rolling red tundra.

A little later the two local bulls climbed a rise overlooking Wonder Lake and descended a short way toward the lake. Then I noticed another group of nine or ten caribou trotting along the lake shore. In the lead was the finest white-maned and spirited caribou bull I have seen. When he noticed the two big bulls above him, he started up the slope toward them, anxious for battle, but before any clashing took place he saw one of my companions and swung about to the lake shore again and continued on the trail. Then he apparently ran into human scent, for he threw his nose upward, looked around, and made a half turn in his uncertainty where to go to avoid danger. His neck and long pendant mane were glistening white, the white extended over his shoulders, and the white lateral line was unusually broad. A splendid bull, already in high rutting fervor! Coming to a decision, he plunged into the lake, and in his wake swam his followers. On the opposite shore they all shook themselves vigorously, a heavy spray of water enveloping each. Then they were hurrying on their way again and disappeared in the tundra. The self-command, boldness of action, and the wild spirit exhibited by the magnificent herd leader fully matched the grandeur and beauty of the radiant fall landscape.

The sublime primitive flight of cranes, the wild and free white-maned caribou bull leading his herd across the lake when land travel seemed dangerous, both incidents taking place in the highly colored autumn tundra, symbolize the

eternal beauty of Alaska. But it is not only the outward
beauty of Alaska that we must think about when considering
its future; we must also think of its native wildness — its
wilderness spirit. This we cannot improve. The problem is to
preserve it.

Alaska still is blessed with many miles of wild country.
Fortunately, the public interest has received some considera-
tion — public forests, bird reservations, and national parks
have been recognized. Still more are needed. True Alaskans
are in love with the rich natural beauty of Alaska and can be
counted on not to emulate the general overdevelopment and
crowding in the forty-eight states to the south.

INDEX

Index

hoofs indicates that they could be used effectively against a coyote or wolverine.

Eagles are known to prey on caribou calves, but the number they kill is insignificant. Not only is the number of eagles extremely small in proportion to the number of calves, but also, so far as I could determine, the eagles seldom prey on young caribou, which are vulnerable for only a short period to eagle attack.

An Alaska Road Commission foreman, Mr. Brown, told me of an interesting incident which occurred about June 13, 1938. He saw a calf left behind by a startled cow. Soon an eagle appeared and commenced to swoop at the calf, which warded off the eagle by rearing up on its hind legs and striking at it with front hoofs. The calf finally took refuge in some willows. When it came forth, two eagles began swooping at it. Once it struck one of the eagles with its forelegs and caused it to cry out. For a few seconds there was a mingling of calf and eagle. The besieged calf again took refuge in the willows, where it lay down. When it emerged, two more eagles had arrived and now four eagles were swooping at it. After fighting them off for a time, it again hid in the willows. The eagles were perched on the ground in the general vicinity of the calf when Mr. Brown left the scene. It was surprising that the calf was able to put up such a spirited defense. If the cow behaved normally, she would in time have returned to her offspring.

I have found five young calves that, judging by the bloodshot marks on them, were killed by eagles. In four cases I did not know the circumstances. The calves might have been ailing or lost. A fifth calf, on which I saw three eagles feeding on July 10, 1948, I had observed the previous day, limping and acting as though sick. When I examined the carcass, I